The supposedly enlightened person's guide to raising a dog

Foreword by Marc Bekoff PhD

Kac Young &
Lisa Tenzin-Dolma

The Hubble & Hattie imprint was launched in 2009, and is named in memory of two very special Westie sisters owned by Veloce's proprietors. Since the first book, many more have been added to the list, all with the same underlying objective: to be of real benefit to the species they cover, at the same time promoting compassion, understanding and respect between all animals (including human ones!) All Hubble & Hattie publications offer ethical, high quality content and presentation, plus great value for money.

More great books from Hubble & Hattie –

Among the Wolves: Memoirs of a wolf handler (Shelbourne)
Animal Grief: How animals mourn (Alderton)
Babies, kids and dogs – creating a safe and harmonious relationship (Fallon & Davenport)
Because this is our home ... the story of a cat's progress (Bowes)
Boating with Buster – the life and times of a barge Beagle (Alderton)
Bonds – Capturing the special relationship that dogs share with their people (Cukuraite & Pais)
Camper vans, ex-pats & Spanish Hounds: from road trip to rescue – the strays of Spain (Coates & Morris)
Cat Speak: recognising & understanding behaviour (Rauth-Widmann)
Charlie – The dog who came in from the wild (Tenzin-Dolma)
Clever dog! Life lessons from the world's most successful animal (O'Meara)
Complete Dog Massage Manual, The – Gentle Dog Care (Robertson)
Detector Dog – A Talking Dogs Scentwork Manual (Mackinnon)
Dieting with my dog: one busy life, two full figures ... and unconditional love (Frezon)
Dinner with Rover: delicious, nutritious meals for you and your dog to share (Paton-Ayre)
Dog Cookies: healthy, allergen-free treat recipes for your dog (Schöps)
Dog-friendly Gardening: creating a safe haven for you and your dog (Bush)
Dog Games – stimulating play to entertain your dog and you (Blenski)
Dog Relax – relaxed dogs, relaxed owners (Pilguj)
Dog Speak: recognising & understanding behaviour (Blenski)
Dogs on Wheels: travelling with your canine companion (Mort)
Emergency First Aid for dogs: at home and away Revised Edition (Bucksch)
Exercising your puppy: a gentle & natural approach – Gentle Dog Care (Robertson & Pope)
For the love of Scout: promises to a small dog (Ison)
Fun and Games for Cats (Seidl)
Gods, ghosts, and black dogs – the fascinating folklore and mythology of dogs (Coren)
Helping minds meet – skills for a better life with your dog (Zulch & Mills)
Home alone – and happy! Essential life skills for preventing separation anxiety in dogs and puppies (Mallatratt)
Hounds who heal – it's a kind of magic (Kent)
Know Your Dog – The guide to a beautiful relationship (Birmelin)
Life skills for puppies – laying the foundation for a loving, lasting relationship (Zuch & Mills)
Living with an Older Dog – Gentle Dog Care (Alderton& Hall)
Miaow! Cats really are nicer than people! (Moore)
Mike&Scrabble – A guide to training your new Human (Dicks & Scrabble)
Mike&Scrabble Too – Further tips on training your Human (Dicks & Scrabble)
My cat has arthritis – but lives life to the full! (Carrick)
My dog has arthritis – but lives life to the full! (Carrick)
My dog has cruciate ligament injury – but lives life to the full! (Haüsler & Friedrich)
My dog has epilepsy – but lives life to the full! (Carrick)
My dog has hip dysplasia – but lives life to the full! (Haüsler & Friedrich)
My dog is blind – but lives life to the full! (Horsky)
My dog is deaf – but lives life to the full! (Willms)
My Dog, my Friend: heart-warming tales of canine companionship from celebrities and other extraordinary people (Gordon)
Ollie and Nina and ... Daft doggy doings! (Sullivan)
No walks? No worries! Maintaining wellbeing in dogs on restricted exercise (Ryan & Zulch)
Partners – Everyday working dogs being heroes every day (Walton)
Smellorama – nose games for dogs (Theby)
The supposedly enlightened person's guide to raising a dog (Young & Tenzin-Dolma)
Swim to recovery: canine hydrotherapy healing – Gentle Dog Care (Wong)
A tale of two horses – a passion for free will teaching (Gregory)
Tara – the terrier who sailed around the world (Forrester)
The Truth about Wolves and Dogs: dispelling the myths of dog training (Shelbourne)
Unleashing the healing power of animals: True stories abut therapy animals – and what they do for us (Preece-Kelly)
Waggy Tails & Wheelchairs (Epp)
Walking the dog: motorway walks for drivers & dogs revised edition (Rees)
When man meets dog – what a difference a dog makes (Blazina)
Winston ... the dog who changed my life (Klute)
The quite very actual adventures of Worzel Wooface (Pickles)
Worzel Wooface: The quite very actual Terribibble Twos (Pickles)
Three quite very actual cheers for Worzel Wooface! (Pickles)
You and Your Border Terrier – The Essential Guide (Alderton)
You and Your Cockapoo – The Essential Guide (Alderton)
Your dog and you – understanding the canine psyche (Garratt)

www.hubbleandhattie.com

First published August 2017 by Veloce Publishing Limited, Veloce House, Parkway Farm Business Park, Middle Farm Way, Poundbury, Dorchester, Dorset, DT1 3AR, England. Fax 01305 250479/email info@hubbleandhattie.com/web www.hubbleandhattie.com ISBN: 978-1-787110-59-5 UPC: 6-36847-01059-1

Readers with ideas for books about animals, or animal-related topics, are invited to write to the editorial director of Veloce Publishing at the above address.
British Library Cataloguing in Publication Data – A catalogue record for this book is available from the British Library. Typesetting, design and page make-up all by Veloce Publishing Ltd on Apple Mac. Printed in India by Replika Press.

Contents

Dedication & Acknowledgements

Dedication

Lisa Tenzin-Dolma

For Skye; a king among dogs; my muse, and the best friend I could have ever wished for.

Kac Young

For Talulah Lake and Truffle, the canines who crept into my heart when I was least expecting it, and brought a world of sunshine, laughter, and new meaning to the term 'best friend.'

Acknowledgements

Lisa

Huge thanks to Kac, whose idea it was to put our heads together and write this book, and who is one of the funniest and wisest people I've ever had the good fortune to meet. Also to Marlene Morris for being there for Kac and me.

To our literary agent, Lisa Hagan, and to Jude Brooks at Hubble and Hattie for believing in this book, and to Dr Marc Bekoff for his wonderful Foreword – I can't thank you all enough.

My children are true blessings in my life. Ryan, Oliver, Dan, Liam, and Amber – thanks for being the beautiful people that you are.

My amazing team at The International School for Canine Psychology & Behaviour cheerfully took on extra work so that I could be free to take time out for writing: June Pennell (the best secretary in the world); Amber Tenzin-Dolma; Andrew Hale; Caroline Wilkinson; Dale McLelland; Deb Lovell; Karen Irwin; Liz Morris; Rachel Hayball; Rachel Parnell; Sarah Western; Theo Stewart; Tom Candy, and Teresa Tyler. You all rock!

I owe thanks to more friends than space allows. Big hugs of love and gratitude (and bottles of wine) to Annie and Bryan Rawlings, Sarah Fisher, Diana Ossana, Amelia Welham, Sue Beech, Toni Shelbourne, Isla Fishburn, Michael Eastwood, Marius Von Brasch, Caroline Wilkinson, Carolyn Menteith, Bel Mooney, Paul Halpern, and Chris Halliday.

Kac

Deepest gratitude to Lisa Tenzin-Dolma for her willingness to hold my hand across the pond while I stumbled my way to learning how to be a good doggy mom. She laughed with me, not at me, as I increased my circle from 'just cats' to cats and dogs.

To Marlene Morris for her patience, support, inspiration, co-parenting, and love throughout this adventure. To my precious agent, Lisa Hagan, for her incredible positive attitude and creative support, and to Jude Brooks at Hubble & Hattie for her brilliant editing, sense of humor, and affirmative support in publishing this book.

To all my friends' dogs who helped me make the big leap: Skye, Charlie, Chloe, Spencer, Marybelle, Rocky, Maggie, Rogan, Runner, Goldie, Gigio, Mr Trouble, Sabrina, Liza, Audrey, Sassy, Yuki, Spike, Vegas, Maya, Chin, Foxie, Chaz, Carlos, Pennywise, and a roster of more that my mind may have forgotten over the years but who left big pawprints in my heart.

To The Rescue Train in Studio City, CA for its dedication, and extraordinary work of rehoming dogs who have been abandoned, surrendered, misused, or just in need of a forever home. Thank you for all the amazing work you do: Lisa Young, Paula Cwikly, Diane Brounstein, Susan Hecht, Noreen Reardon, Heidi Hirsch, and all the volunteers who help our canine friends.

Special thanks to J Randy Taraborrelli (and Spencer), for being the first friend to actually order this book, and for his love and support of me and my work. Thanks so much, dear Randy!

And to all the companion animals who have preceded Talulah and Truffle for their love, understanding, and patience while I learned how to be an adequate caretaker of these life-changing beings. Very special mention to my cats Lucy (who lived for 24 years), and Jabez (my greatest healer).

Foreword
By Marc Bekoff, PhD

Written from two perspectives, that of a first-time dog adopter, and an animal behaviorist, this book is both entertaining and informative. Many people who are new to dog guardianship will be able to relate to Kac Young's experience of adopting a dog after a lifetime as a self-confessed 'cat person.' Through the course of this book we see a shift take place, from Young's impulsive, somewhat haphazard start, to her blossoming understanding of how to make sure that both the canine and human sides of the relationship are in harmony with one another. The stories of her adventures with Talulah, and subsequently Truffle, are interpreted from the dog's perspective by Lisa Tenzin-Dolma, who gives guidance on the needs of dogs, and why it's so extremely important that we pay attention to meet these needs, and give each and every individual the best life possible.

Enlightened guardianship can be defined as taking steps to nurture the human-dog relationship through understanding why dogs behave as they do, knowing how to enrich their lives while gaining their willing cooperation, and through enabling dogs to feel safe, comfortable, confident, and valued as members of their social group: the family.

Dogs have become popular research subjects over recent years, and studies have indicated that the bonds they form with us are intense, enduring, and deeply meaningful. I have written extensively about animal emotions in my books, and in numerous essays for *Psychology Today*, and it has become clear that dogs experience a wide range of emotions, and are sensitive to our feelings and moods. This book rightly stresses that we need to consider their cognitive and emotional lives carefully, even before bringing a dog into our lives, and that their wellbeing is dependent on our own emotional states, attitudes, and behaviour. It's a strongly symbiotic relationship.

The combination of Young's witty, often amusing stories, and Tenzin-Dolma's calm, reassuring guidance, makes this a deceptively easy-to-read book, yet it contains plenty of gems to help new adopters build mutually happy, life-enhancing relationships with the dogs in their care. It is a most timely and significant book that deserves a broad global audience.

Boulder, Colorado
marcbekoff.com

Introduction

Kac

I've been a lover of felines ever since a friend and off-duty fireman hacked through the wall of her apartment to rescue a little black kitten who had been born there. That little kitten was so grateful for life and lived with me for 24 years. Since then, 11 cats have shared my life; currently, I have 6, all over the age of 10.

Then I met Lisa Tenzin-Dolma. Her creativity, her ethics, and her love for dogs made us fast friends. Lisa has always been into dogs, is a leading figure in the animal rights movement in the UK, and founded The International School for Canine Psychology & Behaviour, and The Dog Welfare Alliance. She's co-chair of The Association of INTODogs, and represents this organisation within the Animal Behaviour & Training Council. Lisa is also a member of the Pet Professional Guild, the Dog Rescue Federation, Pet Dog Trainers of Europe, and the National Register of Dog Trainers and Behaviourists.

Lisa has authored four books about dogs: *The Heartbeat at Your Feet: A Practical, Compassionate New Way to Train Your Dog; Dog Training, The Essential Guide; Adopting a Rescue Dog,* and *Charlie – The dog who came in from the wild.*

In our decade of friendship I have fallen madly in love with Lisa's canines, laughed at their antics, prayed for their healing, fretted over their challenges, and reveled at the stories she has shared with me about them. I've been across the pond to meet them twice. Lisa has loads of experience in loving, training, and understanding dogs and their behavior, which is why I asked her to write a book with me about the way to raise and train a dog when you're an 'enlightened' person – meaning a long time student of spirituality and practitioner of all things good to begin with. The starting point makes a big difference.

Having always been a 'cat person,' I surprised myself when I fell in love in an instant and rescued a dog. With Lisa's generous help, we turned this abused little lady into a glorious princess with plenty of panache, and repaired her emotional and physical wounds. One dog led to two, and now I am a cat lover *and* a dog lover, all due to Lisa and her loving instruction and gentle guidance.

This book takes you through the stages of adopting a dog, training him or her, and adapting both of you to a new life with each other. Life changes when you adopt a companion animal, and this book guides you through the steps, alterations, and corrections as you establish a new and loving relationship with a dog.

If you are looking for a book on enlightened, compassionate, and kindness-based dog raising and training, this is the book for you. The best news is that we have Lisa. She's the real star. I'm still a student, but, hopefully, my story and the adventures of my two little canine loves will inspire you to create the best and most perfect relationship you'll ever have with that special creature known as DOG.

Lisa

Scruffy the dog is in one of the earliest baby photos taken of me, and my best friend and confidante at the age of seven was a Border Collie called Bobby, so perhaps it's not too surprising that I love dogs so much. Dogs are wonderful teachers and mirrors. Their unconditional love and devotion, their eagerness to bond with us, even after suffering past abuse, and their ability to live in the moment continually inspires and uplifts me.

Kac and I first met over ten years ago because we had read each other's self-help books, and we instantly became close friends. She's one of the most creative, motivating, and inspirational people I've ever known, and I know a lot of truly amazing people! Even though she was a self-confessed 'cat person,' she became a 'dog person' also after getting to know several of my dogs over the years. The force for change was Charlie, an unsocialized Romanian feral dog who I adopted in February 2013, and whose story is told in my book, *Charlie – the dog who came in from the wild.* Kac visited with our dear friend, Marlene Morris, and it was love at first sight for her and Charlie. This very fearful boy spent every moment smooching with Kac, and offering his paw for her to hold.

Not long after Kac and Marlene returned to the United States I was thrilled to hear that Kac had adopted a small dog called Talulah. Before long a second dog, Truffle, also moved in, and I was informed that Kac's cats were no longer speaking to me. Fortunately, they did eventually decide that canine company was fine by them!

Over the past ten years there's been a revolution in dog training, due to an upsurge of scientific research into dog psychology and behaviour. This has impacted hugely on our understanding of dogs, and has revealed that we have far more in common than was previously thought. Dogs and humans have evolved together over a long period of time. A dog-like skull discovered in Siberia dates back 33,000 years, and DNA analysis of dog remains in China date back 16,000 years. If we take mitochondrial DNA from dogs' maternal lines into account, our history together may stretch as far back as 135,000 years! We have changed the appearance and behaviour of dogs as they became domesticated, but scientists Brian Hare and Vanessa Woods also point out in their book, *The Genius of Dogs* that our canine family members have changed the course of human evolution, too. Dogs truly *are* our best friends.

Those of us who follow a spiritual path, whatever name we may give this, base our interactions with others on loving kindness, compassion, and empathy. These qualities can transform our relationships with the dogs in our lives, as well as with our fellow humans, and this book explains how you can develop bonds of love and trust, and can come to truly know and understand your dog so that you both gain the utmost from your relationship.

1 How a cat person became a dog lover, and other miracles of nature

Kac

Over the years I have adopted cats, and become a willing servant to their every request and whim. My oldest cat was twenty-four when she transitioned across the Rainbow Bridge: even in her last moments she was brave, spunky, and loved life.

In total, I have been the human mother to eleven felines, and godmother to three. For the past forty years they have been my joy, comfort, entertainment, heartbreak, and loves of my life; each one of them claiming a piece of my heart. Currently, I care for six; most of them will pass the ten-year age mark this coming year.

Why, then, would a nice cat lady like me want to own a dog?

The answer is simple. I didn't.

I have dear friends who have dogs, and I enjoyed visiting them, but they were not a conscious part of my life until I got to know them on Facebook. Each of their faces and stories became embedded in my consciousness, and I found myself checking in to see how they were doing with the fever; how they were handling the move, and what next cute trick would be posted. I did all this with plenty of meows in the background.

I met Skye in Bath, in England; then I met *Charlie – The dog who came in from the wild*. I met Spencer Taraborrelli, Marybelle Cook, Sabrina, Audrey, and Liza (with a z) Cadle. I met Mr Trouble Lombardo, Rocky and Maggie Soldinger, Gigio Gambino, and Chloe Burgess-Worsley, They were all very cute, someone-else's-dogs. Beyond that, in my own casa, I was content changing cat litter, and cleaning up occasional hair balls.

Other cat-lover friends of ours, Pamela and Ralph Ventura, invited us to Los Angeles, my former stomping ground, for a pre-Christmas dinner a few years ago. "Oh," they said, "there's also the annual open house on Riverside Drive in Toluca Lake, when all the merchants stay and provide a little pre-Christmas cheer to get everyone in the holiday spirit. It starts at 5:30pm, and we should go before dinner." This sounded like a festive and very holiday-ish idea.

In the metropolis that is LA, Toluca Lake is like a little village: the home of the late Bob Hope, and adjacent to Beautiful Downtown Burbank, the home of Johnny Carson, Jay Leno, Jimmy Fallon, and *The Tonight Show* at NBC Studios. Close by is Universal Studios, CBS, The Columbia

Pictures Ranch, Warner Brothers, and ABC just around the bend. The little village is a mix of show business people and regular folk, yet resembles a hometown street in Anywhere, USA.

So, December 6, there we were at a street fair wearing our Santa hats that we'd gotten in New York City the year before, when we had been all together celebrating the fantastic magical holiday lights in Manhattan.

We stopped at a couple of shops serving cookies and cider, and passed a pet store. Three lovely ladies were out front holding dogs. The dog in the center looked like a mix of small Pomeranian, King Charles Spaniel, and Chihuahua, and was very tiny, shivering with either cold or in fright; I couldn't tell which. She looked at me with her big brown eyes, and I smiled at her, hoping my smile would warm her a bit. The other two dogs, cuddled in red blankets, were pure Chihuahua.

Our group of four 'oohed' and 'aahed' over the little tykes, and I then made the mistake (?) of saying, "Oh, they are soooooo cute. Too bad this one's not up for adoption," meaning the one I had smiled at. "Oh, but they are," rang out a trio of female voices.

I looked at my spouse, and asked, "Do you think ..." Before I could finish the question, there was a nod of agreement (she later told me that the look on my face was pure love ... how could she resist?).

Before I knew it, I'd said "I'll take THAT one."

But, not so fast ... there were applications to fill out, qualifications to be met, and a financial obligation to be agreed to. I needed to be reviewed, interviewed, screened, and visited at home, and then, maybe, they would allow me to adopt this little creature. Holy Croatia! I obviously did not have *any* idea what I was getting into ...

I went next door to the 'open

Hurray! It's Adoption Day. Talulah gets a new mom.

house' at the computer store, and begged ten minutes on the computer. Quickly completing the online application form, I returned to the pet store ... which said it would be in touch ...

"But, but, but, I live 250 miles north," I pleaded, "and I want to take the dog home with me."

It turned out that the three ladies were part of a local organization called The Rescue Train, and they spent hours and hours taking in abandoned pets, rehabbing them, paying vet bills, and then, with the greatest care, finding new homes for them. 'My dog' was being fostered by Paula Cwickly, who lived just around the corner from the pet store (which was providing a showcase for them that night). 'My dog' was so frightened by all the noise and throngs of people, that, by the time I finished filling out the online application, Paula had taken her back to her foster home.

The fun was only just starting. The three ladies (who were pretty sure I had fallen head over paws in love with my little rescue) contacted Lisa, head of The Rescue Train, to tell her of my interest, and the next day I went for an interview with

Paula and Diane Brounstein. It turned out that I and another rescue volunteer had worked together on television shows for Dick Clark Productions, way back when. She knew my name, and also knew me. It also turned out that yet another volunteer had been an intern on the set of *General Hospital* when I produced that show. Then, we discovered that the individual responsible for home visits had been a member of my spouse's congregation in Toluca Lake for a decade, when she was senior minister there. The pieces seemed to be falling together.

After several more conversations, a few personal reference check-up calls, and another meeting, I was awarded 'my dog.' Originally christened Bianca, we decided that, having met her at the Christmas party in Toluca Lake, her name should be Talulah Lake. Even before I picked her up from Paula the morning of December 8, I had already stopped by Petco and bought her a new purple bling collar, and a purple heart-shaped name tag with her new name engraved on it: 'Talulah Lake.'

It was a happy day indeed when I got to write a big check to the rescue, and take home my new little darling. I had bought her a car seat for safety and comfort for the four-hour ride home, and she was a perfect angel, perched in her seat, and eventually lulled to sleep by the rhythm of the road. When we stopped for a bite to eat, I read some of her paperwork, and admit to wiping a small tear from my eye as I learned that this very day was her actual third birthday.

Talulah ate a little bit of her lunch, and accepted a treat. Returning to the car, she jumped up into her car seat, and promptly threw up all over floor. I was thrilled!

Talulah's first steps in her new forever home.

Talulah tried out every position and corner of the rug for napping suitability ...

Lisa

When Kac sent me a wildly excited email, telling me that she had adopted Talulah, and outlining the amazing series of synchronicities that had brought them together, I did a double-take, followed by a happy dance around the living room

... and likewise with every bed she could find.

with my dogs, Skye and Charlie, and then sent a reply in large-lettered, bright pink bold text, with lots of exclamation marks. 'Congratulations!!!!' What could be more wonderful than rescuing a dog you've fallen in love with?

Love at first sight is common, though it can be misleading. A cute appearance doesn't always mean it's going to be a good match, though, fortunately, it was in Kac and Talulah's case! When eyes meet and hearts melt, all logical thought flies out the window. Love is a chemical reaction, and the release of oxytocin, the bonding hormone that surges when we connect with a dog as well as a human, has a powerful effect on both parties. It makes us feel that everything will work out just fine because (cue more dancing) love conquers all. Actually, when you combine love with patience, understanding, commitment, and yet more patience, it really does!

Dogs have an uncanny ability to sniff out good people. Their noses are so much better than ours, and we have to rely more on other senses – intuition included. Skye, my nine-year-old Deerhound-mix boy, is a gentle giant who bestows his friendship as generously on people as other dogs. He's been with me since puppyhood, and his wise, accepting nature has made him a natural muse and mentor to the many other fostered and adopted dogs who have come to live with us over the years. He adores Kac and her spouse, Marlene, so his rapturous greeting when they last visited was just what we'd expected. The big surprise was Charlie's reaction.

Charlie is the subject of my book, *Charlie – The dog who came in from the wild*. Our one-eyed feral boy was sent to me from Romania in February 2013, where he had lived wild all his life. The plan had been to foster him until the right home was found, but it soon became clear that his extreme fear issues would not only make it hard to home him, but that another move would be far too traumatic for him to cope with. Plus, those love chemicals were working overtime! Two weeks after Charlie's arrival I adopted him, and this turned out to be one of the best decisions I've ever made. The long process of helping him adjust to life in a home is described in my book: it was challenging, sometimes heartbreaking, and ultimately incredibly rewarding to live with and love a wild soul – our 'Wolflet,' as everyone called him, because his behavior was at first more wolf-like than dog-like.

Charlie needed time to accept new people into his life, and we had a set of 'Charlie Rules' for visitors, both to keep *them* safe, and to help our feral boy feel comfortable. When Kac and Marlene arrived, however, Charlie's reaction was extraordinary! He danced around them with his tail whirling like a helicopter rotor. He sat in front of Kac and offered a paw, over and over again. He rubbed up against her for smooches. He licked her hands. He made it clear that he really would like to sit on her lap (and Charlie was certainly not a lap dog, except with me). Charm wafted around the room like a magic spell, and we were all entranced by this uncharacteristic behaviour.

When I heard that Kac had adopted Talulah, somehow, I wasn't too surprised, though I *was* thrilled. I had a sneaking feeling that somehow Charlie had been involved in this!

The adoption process may seem arduous for first-time adopters, but there are good reasons for all of the checks. Sadly, sometimes adoptions fail: either because love at first sight turned out to be not what it seemed, and infatuation doesn't last; sometimes because the match is all wrong, and often because it can be hard to accurately assess dogs in kennels (that's why fostering is so important), and it can take at least a few weeks – sometimes months – for a dog to properly settle and unpack his or her baggage from the past to reveal their true personality.

Talulah experimented with different locations in the house, and different beds until she found the perfect one. This is it.

How's that one working out for you, Talulah?

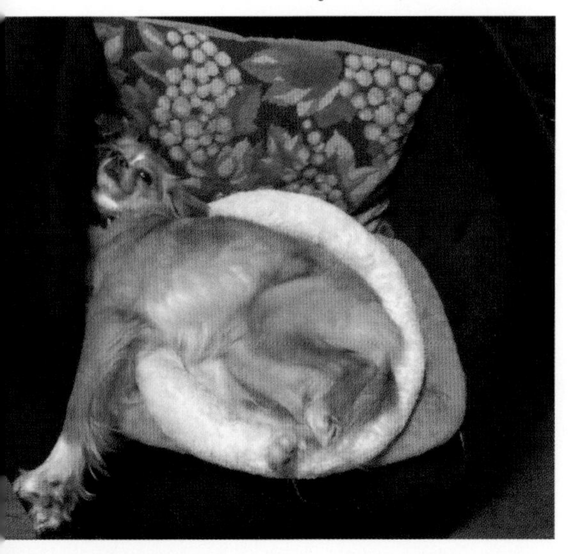

Questionnaires need to be filled in so that the rescue centre can gain some insight into the adopter's circumstances and lifestyle, and a home check is carried out to ascertain whether boundaries are safe, and if the prospective adopter is going to get antsy about dog hair on the rug, or the occasional accident during toilet-training. Some rescues ask for references, to ensure an individual has what it takes to properly provide for a dog's physical, mental, and emotional needs, and also to ensure there's no intention to sell-on the dog, or use her as a breeding machine if she hasn't been spayed.

All of this is also for the adopter's benefit. It's heartbreaking to adopt a dog and then find that it's not working out, and the dog has to be returned to the shelter. The emotional impact when this happens is hard on the human, but even tougher on the dog, who doesn't understand why he or she is suddenly homeless once again. It may also be harder to find him an alternative home because he has been returned. Even if it's not his fault it can give him a bad rap.

Getting the all-clear that the adoption is going ahead is a hugely exciting time! As well as the practical aspect of buying all the necessary accoutrements (collar, harness, leash, dog bowls and beds, food, toys, treats, a crate, and anything else that you feel is a must-have), there are plans to be made, people to be told, and all the anticipation of and questions about life with your new family member.

How will your dog travel – in a car seat or a crate in the back of the car? Will he or she settle in quickly, or will there be some glitches whilst you get to know each other?

When you first bring home your dog, it can help to remember that, although *you* know what a great life your dog will have, your new furry family member has no idea what's going on. She's going to be confused, scared, and disoriented, and won't know your house rules, or whether she can trust you to treat her kindly. Everything will be unfamiliar to her – the environment, the scents and sounds, the food (it's always best to check what she has been eating, to make the transition smoother and avoid upset tummies), and the areas where she's expected to eat and to relieve herself.

All things considered, it's amazing that most dogs manage to adjust pretty smoothly!

How to find the right dog for you

We all have different lifestyles, so if you're very active and want a walking or running companion, an energetic dog will be a better match than if you adopt a couch potato who wants to sleep most of the time. If you prefer a quiet, fairly sedentary life, you won't want a high-energy dog who may disrupt that.

If you're planning to adopt a rescue dog, make sure that the rescue carries out thorough assessments to ascertain how the dogs in its care cope with children, other dogs, and cats, bearing in mind that a dog may respond differently in a home to how they might in a shelter. If assessments aren't done, you could ask a qualified, positive, force-free trainer or behaviourist to go with you to meet the dog.

The following tips will help you to decide how to choose your new companion.

1 Think about why you want a dog. Are you looking for a companion to snuggle up with on the sofa, or a dog who will grow up with your children? A jogging companion and playmate, maybe, or a dog who you can teach tricks to?

What about a working dog? Please note that service dogs are another matter altogether, as these are specially trained to carry out specific tasks, and should come from a reputable organization that has carefully matched the animal with the new handler.

2 How about your lifestyle? Are you immensely sociable, love going out with friends and having visitors? In which case, your dog should be one who enjoys having lots going on. Are you quiet and reserved, and prefer to stretch out on the couch at the end of the day to relax with a book or spot of tv? A chilled-out canine would be ideal for this, rather than one who wants fun, fun, fun!

3 What age should the dog be? Puppies are great fun, of course, and you can shape them to an extent as they grow up, but they can also be hard work as you need to teach them house manners, and ease them through the tricky adolescent

stage when dogs (like teenagers) tend to test boundaries.

Many puppies can be found in shelters, but, if you plan to buy an animal from a breeder, do your research first to ensure the pup has had the best possible start in life, with positive exposure to people and other dogs. Puppy mills are rife, sadly, and are responsible for immense suffering for both parent dogs and their puppies. Meeting the dam, and visiting her in her home environment, is essential when considering buying a puppy.

Older dogs have usually already learned house manners, and tend to settle in well in their new homes, though some arrive with baggage from difficult pasts, which is why it's so important to gather as much information as possible before adopting.

4 Research dog breeds. All dogs have unique personalities, but certain traits are predominant in certain breeds. Knowing as much as possible about the breed (or mix) that appeals to you will help ensure that the dog will be a good fit with your home and lifestyle.

Generally, the larger the dog, the shorter his lifespan, so this is something to consider, although this should be considered a long-term commitment, whether you adopt a Great Dane, who may only be with you for six to eight years, or a Jack Russell Terrier, who you may have for seventeen or more years.

The main points to look for are –

- General temperament
- Activity levels
- Trainability and sociability with people and other dogs

5 Ask questions. Discover as much as possible about how the dog you're attracted to gets on with people, children, other dogs, and small furries such as cats and rabbits. Does he have any known issues, such as pulling on the leash, fear, aggression, or food-guarding? Does he enjoy being stroked, or is he aloof, preferring to receive affection on his terms?

Not all rescue dogs come with a known background, but it's still possible to determine how he or she responds to strangers and other dogs. It's important to bear in mind that kennels are very stressful places, and a dog who isn't coping well in a shelter may take to home life like a duck to water. It can also be useful to remember that a dog who has spent years in a shelter may have become institutionalized, in much the same way humans can, and may need time to adjust to life in a home.

6 Consider costs. Buying a puppy from a breeder, or paying a donation fee to a rescue shelter is just the beginning of the costs involved. As mentioned previously, your new family member will need a bed, food and water bowls, food, treats, a collar, harness and leash, perhaps a crate, and (with dog breeds who feel the cold) a dog sweater or coat. You'll also need to factor in the cost of lifelong veterinary care, vaccinations, and treatment. You may choose to take your companion to training classes, or enlist the help of a one-to-one trainer or behaviourist. Dogs don't come cheap, so be sure you can afford to meet all of your new friend's needs.

7 Pet insurance. I can't emphasize strongly enough how vital it is to arrange to have this in place when you bring home your dog, because, if he should become sick before you do so, any insurer you subsequently go to is likely to view this illness as a pre-existing condition, and you'll be liable for the full cost of future medical care for anything related to it. Medical costs tend to be high, especially if your dog needs scans or x-rays, surgery or long-term medication, but if you have insurance you'll only have to pay a small portion of the veterinary fees.

Shop around before you get your dog, as insurance premiums and coverage vary. You'll find you have the choice of yearly insurance or insurance for life, and, in my opinion, the latter option is best because if your dog becomes sick, the illness will be covered for his or her remaining years. Yearly insurance policies subsequently regard any illness during that year as a pre-existing condition, and exclude it from what's covered.

8 Pre-adoption supplies. It will be much better for you and your new dog if your home is already prepared for her arrival, than if you collect her and then have to rush around a pet store, gathering all she needs. Take a collar, harness and leash with you when you collect her, and remember to attach an identity tag engraved with your name, address, and phone number to her collar, but not her name. If she became lost, say, knowing her name would be helpful for someone with good intentions, but a person with less noble intentions could use the information to entice your pup away.

Have a dog bed ready in the area where you want her to sleep, and place food and water bowls of an appropriate size for her breed in her feeding area. If using a crate, line this with a soft cushion or blankets to make it cosy, and place a soft toy or filled Kong™ in it to make it appealing. Leave out just one or two dog toys at a time (occasionally swap these around to make them more interesting), and make sure you have lots of small, tasty treats for rewarding all of the behaviours you want her to repeat.

9 Diet. A good quality, appropriate diet is the first step towards keeping your dog healthy in the long term. Ask the rescue staff which food your dog has been eating, and buy enough of this to last for at least the first week. The stress of moving home can result in upset tummies, and a change of diet won't help.

Decide whether you want to feed a raw diet, pre-cooked, ready-prepared food, home-cooked, or complete dried food. If you research dog foods and talk with your vet about the pros and cons of different diets, you can make an informed decision. When you change the diet, do this gradually over a period of a week or two, beginning with mixing just a little of the new food with her current diet, and gradually substituting more until you are happy she likes it, and the changeover is complete.

10 Settling in. The settling-in process can be helped immeasurably by taking things slowly, giving your dog time to find her way around, and get to know you and her new home. Nothing creates more stress for a dog at an already scary time than being surrounded by lots of strangers in an unfamiliar environment, so although you'll naturally want to introduce her to your family and friends, and celebrate her arrival, please hold back on this! She'll really appreciate being given a week or two in which to get used to you and her surroundings, without the additional pressure of visitors wanting to make a fuss of her.

Let her find her way around and get to know you. Avoid putting pressure on her to interact, and allow her to choose when to come to you for a stroke. Reward all desirable behaviours with a treat and praise, so that she learns what you want from her. If she's doing something you don't want, such as chewing something she shouldn't, please avoid chastising her: instead, kindly call her to you and give her something appropriate to chew on.

11 Teaching her name – and that all-important recall. Begin teaching your dog how to respond to her name as soon as she arrives, and teach good recall at the same time by calling her name from across the room, and from other rooms, rewarding her handsomely each time she

comes to you. Keep your tone of voice light and happy to make coming to you more inviting, and give her a tasty food reward as soon as she does. Even if she becomes distracted en route and takes her time, the reward acts as an incentive, and she'll soon learn to respond more quickly: invaluable when you begin to take her out for walks!

12 Toilet training. This should be started straight away so that good habits form quickly, though it's only reasonable to expect some accidents in the early days. If you have a yard or garden, take her through your home and outside on-leash as soon as you arrive, and give her time to sniff around. If she 'performs,' immediately drop a treat in front of her, and tell her, in a happy tone of voice, how good she is. If she doesn't want to go, simply take her back indoors and let her explore for a while before returning outside. Stay with her each time you take her out, so that you can mark any toileting with goodies and praise on the spot. This will teach her that she's doing the right thing.

Indoors, be on the alert for any signs that she needs to 'go' (sniffing, circling and looking restless), and immediately act on these. Also take her outside after naps, playtimes, and eating and drinking, as these are the times she's most likely to need to eliminate.

Refrain from getting upset if she has an accident. Remember that she's learning, and quietly, calmly clean up. If you tell her off she's likely to become anxious and look for places to hide when she needs to do her business! Toilet training is easy if you're calm and consistent, and it quickly becomes a new habit.

Summary

- Appearances aren't everything, though love at first sight *does* happen
- Your lifestyle guides the choice of the perfect dog for you
- Checking out dog breeds and considering whether a puppy or older dog will suit you best will help refine your choices
- The more information you have about a dog's background, the easier it is to ascertain whether the match is right
- Be prepared for the not inconsiderable expense involved in caring for a dog
- Arrange for pet insurance to be in place as soon as you bring home your dog
- Look into options about the best diet for your dog
- Give your dog time and space to settle in, then start immediately with teaching recall and toilet training

Talulah moved from southern to central California, where the temperature dropped 20 degrees. Her new puffy coat helped her adjust to colder weather.

2 Tu-Rah-Lu-Tah-Lu-Lah

Kac

I couldn't help myself. I was *so* excited about being able to take my new baby girl with me on outings, I bought a pet screen partition for the back of the Jeep, a new, gold-colored doggie bed, and special no-tip traveling water and food dishes for Talulah. I wanted her to feel safe and comfortable when I took her with me on excursions and errands. I had read that it is much safer if an animal is restrained and placed in a safe compartment while traveling.

The first day I was very excited to show Talulah her 'traveling quarters.' She took one look and wanted none of it. "No, I, Princess Talulah, want to sit on your lap while we go in the car," she seemed to tell me with her big brown eyes, and the way she cocked her head to the right. I reasoned with her in our driveway, until she reluctantly agreed to do it 'my way.' We headed out and she began to fuss. All the way to the store and back she cried, shivered, and wailed. I stopped the car twice to see what was going on, and although she appeared to be safe and fine, the whining did not stop. I was frustrated because I felt I had done all the right things for her, and yet she was miserable.

Talulah can watch the world go by from her dog seat in the back of the Jeep.

I rationalized that this behavior might just be due to the 'newness' of everything, and she would feel calmer in time and become a better traveler. She might even grow to like it, I fantasized.

With optimism in my heart, Marlene and I took her on another journey, four times longer than the first trip, along the beautiful Pacific Ocean coast. Talulah did not appreciate the sweeping views, however, and whined, shivered, and cried all the way along that exquisite shoreline. This time I had a witness, and we looked at each other in desperation.

I have had some success singing to my cats when they are fussy and agitated, so I suggested we sing her a song, and made one up on the spot. There's nothing like a good Irish song to brighten the day, and so, taking a few liberties, we sang our version of a familiar Irish tune –

Tu-rah-lu-tah-lu-lah. Tu-rah-lu-ra-lie,
Tu-rah-lu-tah-lu-lah, hush now don't you cry.
Tu-rah-lu-tah-lu-lah, Tu-rah-lu-ra-lie,
Tu-rah-lu-tah-lu-lah, it's a daw-aw-ggie lull-la-bye.

We sang it repeatedly; we sang it in two-part harmony, and we sang for forty minutes straight. It kind of worked for a brief moment but didn't solve the problem by any means. By the time we got home, Marlene and I felt like we had auditioned for a reality singing show ... and lost.

I emailed the ever-smart Lisa and described my woes. Here is what she suggested ...

Lisa

While some dogs love riding in cars (especially if they're able to poke their noses through a small opening in the window to snuffle all those scents they're whizzing past), others find travelling horribly uncomfortable. If your dog is an anxious traveller, it could be due to motion nausea, and, of course, she won't understand why the world is passing by at a frighteningly blurry speed, whilst she's being shaken and stirred by bumps and turns in the road.

One way to relieve this is to cover the dog crate with a large cloth or sheet, so that she can't see all that is going on, and, often, she'll settle and fall asleep. You can also try putting a stuffed toy, or worn, unwashed item of your clothing in with her for comfort, and add a stuffed Kong™ filled with light, tasty nibbles to keep her happily occupied. Dog chews aren't a good idea in cars, because it's hard to supervise and make sure your dog doesn't choke, so a Kong™ filled with tiny treats, pieces of cheese or chopped chicken is best. You can even prepare several and freeze them, ready to offer on long journeys.

If your dog is wearing a harness clip on a car seat, you can use opaque static window film on the side windows to block the shifting views. You can find this in DIY stores, and all you need to do is cut window-sized sections from the roll, dampen the glass with a wet cloth, and smooth on the film. It peels off easily, leaving no marks, when you no longer need it. This is also very useful indoors for dogs who bark and become over-excited or over-aroused should people pass by.

Bach Flower remedies can help a lot. Rescue Remedy is great for easing stress and anxiety, and Scleranthus can help to relieve motion sickness. Just add two drops of each to food and water twice daily if you're planning regular journeys. Before leaving home, put a drop of each remedy on your finger, and stroke that onto the top of your dog's head where the dome shape can be felt.

If your dog doesn't like travelling, take it slowly at first, and make the car a good place to be in. I've worked with rescue dogs who, having actually been thrown out of moving cars, are justifiably phobic about going anywhere near anything with four wheels.

Some dogs simply associate being in a car with feeling uncomfortable or distressed, and the best way to work with this is to create positive feelings about cars before taking your dog on a drive. Here's how to help her overcome car aversion.

Each mealtime, take her dish and place it on the ground a short distance from the car (it can help if you include in her meal a special little something that you know your dog particularly loves). All of my dogs would happily move mountains for a piece of cheese, for example, but yours may prefer chicken, or peanut butter, or a few slices of sausage, salmon, or a little pate mixed into the food. If she won't approach the food, this means the bowl is too close to the car, so move it further away. Give fulsome praise if she eats even only a little. Like us, dogs can't eat when they feel stressed, because a churning, nervous stomach doesn't want food anywhere near it; if your dog eats, that's a good sign!

Gradually move the food bowl closer to the car, and, in-between mealtimes, play her favourite games nearby, too, making sure not to put her under pressure to get too close to the car, and end the game while she's still having fun so that she's eager to return next time.

As she becomes more relaxed, open the car door and feed her just outside of it, on the ground, slowly graduating to the next big step of feeding meals *inside* the car, on the seat or in the back (wherever she will be travelling in future), with the door open. Don't turn on the engine just yet, though! The next step is to give meals inside the car with the door closed, and you sitting with her. Once she's completely comfortable with that, turn on the ignition and drive just a few yards before returning. Again, if she's okay with that, increase the length of time she spends in the car, but remember not to rush into heading out on a long drive.

Talking bodies

How can you know when your dog is anxious or afraid, when you don't speak the same language? Well, actually, she's constantly telling you how she's feeling!

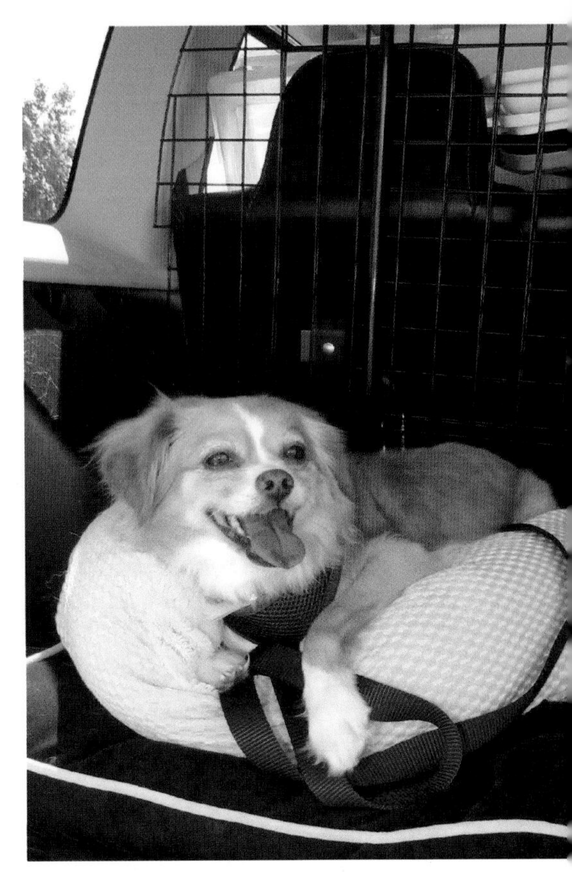

Talulah now loves riding in the car in her very cozy and comfy dog world ...

Dogs communicate by vocalizing, by scent (all that sniffing around gives them so much information!), and by touch, but their main method of expressing themselves is through body language. Anyone can learn to talk dog: you just need to know what to look for, and pay close attention.

An anxious or scared dog draws her body inwards to make herself appear smaller. Her back rounds and dips, her tail drops and may be tucked between her legs, her head lowers and she'll look away from the object of fear. Her ears flatten. She may yawn when she isn't tired, or lick her lips or nose. She'll either freeze in place or try to escape, and she may flatten to the ground. If she's very scared, she may roll onto her back, and even urinate.

If your dog is showing any of these signs near the car, or in any other situation, you need to act immediately to move her away from whatever is causing her stress. After all, you're her guardian and protector; you hold all the keys to her happiness, comfort, and wellbeing, and if you show her that you won't put her in a situation that is overwhelming for her, this will build trust between you. She'll learn that you wouldn't do anything to compromise her safety, and she'll be more willing, and even eager, to follow you to the ends of the earth – or at least, eventually, into the car!

Summary

- Although some dogs love travelling, others become very anxious in cars
- Think about why your dog dislikes car rides. Could it be motion sickness, anxiety over the world passing by too fast, or a previous bad experience?
- Consider remedies for motion sickness, or cover her crate or the window to screen passing views
- Give extra-special meals in the vicinity of the car, gradually decreasing the distance
- Note your dog's body language so that you can tell when she feels worried, scared, comfortable, and happy

... and there's room for two!

3 Twinkle Twinkle, Little Star

Kac

It would never have occurred to me that where and when a dog relieves herself would become a huge issue. I somehow believed that a creature as elegant and delicate as Talulah would not have the common urges of a larger canine, or, say, a racehorse. In this I was deluded.

I had not prepared my brain for frequent trips outside. Before acquiring the lovely Talulah, I figured I would just take her on 'walks;' she would discretely do her business, and we would return inside. The three to four minute break would be good for us, get me away from the computer for a bit, and provide us both with some fresh air. This, I figured, would happen a couple of times a day: morning and night.

Another misconception concerned the deck out back, and I even thought maybe a wee-wee pad could be a good thing for quick potty breaks, or for when nature called when I was on the phone.

Talulah loved to go outside. As a friend laughingly told me years ago, "A dog sniffs the ground like they are reading a newspaper." Ground-sniffing and information-gathering quickly became an hour-long adventure, and Talulah looked like she was leading a parade when we took our walks. I was of a mind to quickly get the harness and leash on Talulah, take a brief walk outside, and then get back to work, but Talulah had other ideas. For her, walks were perfect opportunities to enjoy a stroll down the Champs Elysee whilst modeling her Easter Bonnet. 'Yikes. I can't do this four times a day! I'll never get my work done,' was the thought that swirled through my cranial cells.

I racked my brain. I thought that if I put out a grass mat, especially made for dogs, this would be 'the spot' Talulah would want to do her business, and then quickly return inside. Wrong! Talulah veered away from the grass pad and headed off down the deck as if I had insulted her entire family. She turned away from me, jumped up into a pot and used the dirt underneath the Fichus tree as her bathroom. I heard a low growl, and realized it was me making the sound, and not her.

For a while, we braved the unfenced front yard, walked a chosen path, and took as much time as Talulah needed for the doing of her business (which was usually much longer than I wanted to be out). Some days, Talulah would be nature-girl and just stand in one place smelling the air and listening to the birdies chirp, living a Disney movie. I could picture a Monarch butterfly alighting on her nose and Talulah breaking into a dance number. This was all charming and wonderful for a Sunday afternoon frolic, but when deadlines are looming and it's a working Tuesday morning, something had to change, or mommy would be out of a job and living under a freeway with her new puppy in a cardboard box.

I emailed Lisa about my latest challenge. She suggested I find one word and use it repeatedly so that Talulah would recognize it, want to please me, and go about her business in a more time-sensitive fashion. Walks would now have to be completely different and the pace would not be as rapid as some of the potty breaks needed to be. It was my job to make the distinction so she knew the difference. Alright, so now I'm Dr Phil ...

Before trying this, I thought maybe I could outsmart Talulah.

I purchased a large frame dog kennel which measured five feet by ten feet and six feet high. It had a front gate, and positioning it out on the back deck made sense so Talulah could be out there all by herself, romp in the sun, listen to the birds, and sniff the air to her heart's content. I placed her doggie-potty pad in the kennel for her convenience.

How wrong can you be! Once again, well-intentioned but clueless mommy messed up. Talulah hated being in the big cage all by herself, and whined and cried every time I put her outside to commune with nature. She wanted to be with me, or at least in close proximity, rather than outside talking to a Blue-jay.

I decided to go with Lisa's advice.

"Pick a word," she had told me, "and use it."

I chose the word 'twinkle,' and, to this day, Talulah 'twinkles' on cue. It didn't happen overnight, mind you, but as I summoned more patience, she came to associate the word with the action of having a quick wee. Happy Mommy; happy Talulah!

Talulah learning to 'twinkle' outside in the garden ...

There's more about the twinkle word later on in the book, because the story isn't over yet. For now, though, suffice to say that 'twinkle' was the breakthrough keyword to much more effective training and good results.

Lisa

Toilet training issues are just one of the many things that land dogs in shelters, yet it's really not difficult to teach your dog to be clean indoors. A lot of the dogs who have come to stay with me hadn't lived in a home before, either because they were born wild, like feral boy Charlie, lived on the streets, like my Greek foster dog, Saffy, or had spent their lives in kennels, as had most of my ex-racing Greyhound fostered and adopted dogs. They all needed toilet training, and the longest it took for any of

them to 'get it' was just three days. The tools we need are patience, consistency, vigilance, and the ability to move fast if we see any signs that the dog needs to go outside.

When Kac told me that she'd expected Talulah to need just a couple of trips out a day, I imagined how tough it would be if *we* were only allowed two comfort breaks a day: we wouldn't be house-trained, either!

Kac's use of the word 'twinkle,' combined with frequent breaks, demonstrates just how simple toilet training can be.

If you've had a child, you'll remember the potty-training stage well, I imagine. There's the introduction to the potty, the encouragement, the celebration when something lands in the pot ... and the need to sit your toddler on that pot frequently, and stay close by to avoid accidents. It works in exactly the same way with dogs.

Take your dog outside frequently, and always after drinking, eating, sleeping, and playing. These are the times she'll most likely need to 'go.' Look out for the telltale signs of restlessness, circling, sniffing the ground or whining, all of which indicate that she needs to go outside immediately!

If there's an accident, please refrain from chastising her. Remember that she hasn't yet learned what you want from her, and it isn't her fault. If you tell her off, she may find places to hide away to do her business out of fear. Some dogs get so anxious from being repeatedly punished for eliminating that they'll go to the extreme of eating their faeces (in science terms, this is called coprophagia) to avoid being discovered.

Simply clean up without fuss, using either a designated cleaner from the pet store or (much cheaper) biological washing liquid. This removes all traces of the scent, and makes it less likely that your dog will think this is a permitted toilet area.

... job accomplished, Talulah looks suitably pleased with herself.

Bleach won't work. It covers the smell for us, but not for dogs.

Like us, dogs are social creatures, who don't like to be alone, hence Talulah's indignant protests over her outdoor kennel. During the thousands of years we've spent together, we've ensured that dogs have become dependent on us, so it's no surprise that they find it hard to be deprived of our company. Another issue that leads to dogs being relinquished to shelters is separation anxiety, which is terribly distressing for both the dog and their guardian. So, being close by our dogs, ready (with a pocketful of tasty treats) to reward every pee or poo, is the key to effective toilet training.

Follow that nose!

Dogs like to sniff around, and our idea of a walk is very different to theirs. It's tempting to set a route and march from A to B, then home again, when we take out our dogs, but this isn't much fun for our furry friends. They gain a vast quantity of information from their highly sophisticated noses, and use scent to make sense of the world around them. We have five million scent receptors in our brains, whereas dogs have between 125 million (a Dachshund) and 300 million (a Bloodhound). The area of your dog's brain that's designed to identify smells is forty times larger than our olfactory area, so you can imagine just how important dogs' noses are to them! And you can capitalize on this while teaching toilet training.

Your dog uses her nose to gather information about who has been where, what they've eaten, their state of health, whether they are fertile, and numerous other fascinating details – such as whether a flower patch has been visited by bees or ants. If you've been out without her, you'll find that, returning home, especially after visiting friends who have animals, your dog will have a good sniff of your hands and clothes to determine whose company you were keeping.

You can choose an area of your garden as a dog toilet, and take her there regularly (on-leash, in the early days, if you have just a small designated toilet patch in a large outdoor space), but when you take her on walks the wonders of your dog's olfactory sense can be made good use of. When you go out, give her lots of time and space to use her nose. She'll enjoy picking up 'pee-mails' left by other dogs, as these are chock-full of information about the previous passer-by's state of health, age, wellbeing, and sexual availability. You'll notice that your dog has a good sniff around, gathering information, and then peeing close to (or on top of) the previous message, to leave her own mark for other dogs who pass that way. If you take your dog outside, especially to areas where other dogs visit, she'll find the temptation to leave her own messages impossible to resist. As soon as she starts to 'go,' use your marker word (for Talulah, it's 'twinkle,' but you can choose any word that suits you), and drop a treat on the ground right in front of her nose. This instant positive feedback has the effect of making sure she understands that she's done the right thing.

It has to be instant, though! If you're some distance away (let's say it's raining and you want to shelter close by while your furry friend ambles around, sniffing), you're not in a position to immediately say 'Good girl!' and drop a treat while she's squatting, or he's cocking his leg. It can all go wrong if you call her to you to gain her reward, because you're then reinforcing recall, instead of 'twinkling'!

Summary

- Toilet training is easy – when you know how!
- Choose a word that your dog will learn to associate with 'Go toilet'
- Be reasonable about how often your dog needs to go outside: frequently is best
- Take the same approach as you would when potty training a toddler
- Take her out after food, water, naps, and playtimes
- Signs such as circling, sniffing, pacing or whining mean your dog needs to go outside
- If accidents happen, clean up calmly and quietly
- Stay close to your dog so you can reward her as soon as she 'goes'
- Let her sniff around outside to pick up pee-mails from other dogs ... and leave some of her own
- Give lots of positive feedback. Celebrate every outside achievement!

4 Girl gone wild

Kac

The first few days of having Talulah in our home were full of new and interesting discoveries. Deciding on a good location for her food and water dish was an enormous creative hurdle, for example. The six cats who ruled the roost were used to having their electric spring stainless steel fountain all to themselves, and they had free roam of the sizeable house with plenty of seats, furniture, and cat trees to drape themselves on and over.

One of the first orders of business was to purchase a 'baby gate' so I could sequester the den for Talulah, since the six cats had decided not to be members of the welcoming committee. As if I had invited Satan to come live with us, they glared at Talulah with wide, purposely exaggerated stares, as though she was some kind of pop-fiction maniac chainsaw killer coming after them. Not one of the felines greeted Talulah with a hint of etiquette. There were hisses and silent abhorrence, and privately they gave me 'the look' that said it all. I had betrayed them; this was treason. If this had been France in 1793 I would have been next in line for the guillotine after Marie Antoinette.

Our once happy and contented

Kitty-land. For the safety of canines and felines alike, we created a gated community for the cats.

home had become a country divided, but the 'baby gate' would resolve territorial problems for the period of adjustment we would all have to endure for the next few months. That – and a few bottles of Merlot – just might take care of business.

Talulah resided in 'puppy-land,' which comprised the den, kitchen, office,

Percival enjoying dinner in kitty-land.

and sunroom. The cats lived in the rest of the house – three bedrooms, two bathrooms, and a large living room – in which they could romp, cavort, and nap. This came to be known as 'kitty-land.'

At the suggestion of my friend, Hadley, who had rescued a beautiful Coton De Tulear named Angel, I trained Talulah to sleep in her crate at night. What with kitty-land and puppy-land and the great divide, it seemed smarter to let the cats sleep with us in the bedroom, with Talulah comfortable and safe in her own bed.

Talulah had six full days of getting acquainted with her new home and surroundings before we had to leave her alone while we went to church on Sunday. The first Sunday we were to be gone for two hours, and I thought it would be a good idea to leave Talulah in her crate bed while we were out of the house. I didn't know which creature might try to scale the walls to their neighbour's territory, and wanted to employ every caution. I didn't know if Talulah might eat the couches, tear down the drapes, or upend the crystal cabinet while we were away, so I did the most sensible thing I could think of: I put her in her crate, patted her on her little head, and went off to church.

Immediately, the whining began. Before we were out the driveway we could hear her shrieks and cries, as if she was being tortured. I knew she was safe, so we went ahead to church, since we were responsible for the service and couldn't be late. Too many people depended on us being there. But I was worried.

During the church service, I tried to keep my attention in the moment, but my thoughts and feelings were with Talulah. When we got home, we found that she had eaten her puppy bed, as well as pulled a nearby suitcase strap into the cage and chewed the ends off it. She was hot, sweaty, and wild.

'What have we done?' I agonised. Immediately, I released Talulah from her crate and she instantly calmed down. Poor little thing. She plunged her face into the water fountain and panted. It wasn't hot inside the house, we did not live in a tropical climate, but she was overheated and stressed-out. It made no sense to me. I just didn't understand why she was so upset.

On reflection, I thought maybe it was due to having been transported in cages and crates to shelter events; maybe being kept in a cage with other dogs; maybe her previous owners incarcerated her for long periods of time. It was impossible to know for sure, but I knew this leaving-her-at-home-in-the-crate thing was going to be a challenge.

The next few weeks we took her in the car when we went to church. It was cool enough in December, but not too cold, so she waited in the car while we were in church, and got to go for a walk afterwards as a reward, plus a frozen yogurt with mom as an added treat for being a good girl, and not destroying the inside of the car. Although this tactic seemed to be working, we also needed to train her to be able to stay alone for a few hours at a time, so we could go to a movie, or a play, or a meeting, or out to dinner with friends, and not have to worry about her freaking out. I love Talulah, but I was not willing to have my life be all about her, all of the time.

Lisa advised me to take some items of clothing that I had worn, and leave them in Talulah's crate with her so she would sense my presence. (I hadn't replaced her bed, waiting to see if we could fix the problem before investing in another accessory that might also become dinner.) As an experiment – and hoping for the best – I layered towels in the bottom of her crate, threw in some worn t-shirts, and partly covered her crate with a baby blanket.

The next time she was left alone she behaved better. There was no excessive whining until she heard us drive in, although she did manage to chew up the baby blanket while completely ignoring the toys and treats I had left for her. The small water dish was also upside-down. It wasn't perfect but it was better.

What was going on? Was it separation anxiety? Was it just doggy craziness? I called on Lisa again for advice.

Lisa

Kac's use of a safety gate is an excellent idea. I often recommend this when fosterers and adopters bring a new dog into the home, as it enables all of the animals there to see, smell and hear each other, and become used to the extra company safely. Adding a dog to a family of cats is a situation that requires vigilance and careful monitoring, so separation on different sides of a gate means you can relax a little. Once the co-habitants become familiar with each other, allow more up-close and personal introductions.

A dog crate should be a place of safety and reassurance for your dog, to which she can retreat when she feels overwhelmed, or simply needs a nap in a quiet space. In the wild, dogs dig dens in which to stay cool in summer and warm in winter (Charlie, my feral dog, created impressive dens in my garden), and a crate can be regarded as the domestic equivalent. Crates should never, ever be used as punishment; you'll want your dog to view her crate as her sanctuary.

Crate training is a straightforward process. Firstly, make sure it's big enough for your dog to comfortably stand up and turn around in, but not so large that there's space to roam. Add a comfortable dog bed, a bowl of water, and some tasty treats. An unwashed item of your clothing can make the crate feel familiar and less scary for a dog who's unused to confinement. A Kong™ stuffed with goodies can help your dog see the crate as a good place to be, and it takes longer to excavate a Kong™ than to snaffle up a few treats.

However, a stressed dog is rarely interested in food, so it's important that the crate is viewed as a pleasant place to be in its own right. Encourage your dog to enter the crate by putting the food rewards inside, and make sure to leave the door open so that she can go in and out. Once your dog is happy to go inside, you can close the door for just a moment before opening it again, leaving the door closed for a few seconds longer each time you do this. When you use this method, most dogs will choose to sleep in their crates.

Talulah's distress at being left in her closed crate for two hours was due to her not being used to being inside with the door closed during the day. This is where it can help to build the time gradually, leaving the room for a few moments, then returning and opening the crate door without making a song and dance about it. The time in the crate can be extended a little each day, so that it becomes no big deal when you do eventually go out for longer periods.

Talulah's whining and destructive behaviour was due to a combination of separation anxiety ('Help! My Mom's left me and may never return!'), and feeling trapped in a space that she hadn't yet had time to grow used to being in without Kac's reassuring presence nearby. She would have been panicking that Kac had abandoned her forever.

When dogs are stressed they pant a lot, and that makes them very thirsty, hence Talulah immersing her head in the water fountain.

Kac

What we eventually concluded is that Talulah didn't like being left alone in the crate during the daytime, and was perfectly fine, no destruction at all, when we left her in puppy-land, behind the gate in the den area. I left worn t-shirts for her in an open doggy bed, and a nice chew bone. The first time I tried leaving her out in the open I fully expected to come home to shredded furniture. There wasn't a hint of destruction. She was fine.

The crate is still used as the preferred night-time doggy bed but not for daytime seclusion. Talulah can be calm, happy, and quiet for hours and hours on end. She hasn't eaten another baby blanket since the first episode, and we are confident we can leave her alone for up to five hours maximum with no negative results. We do leave a potty pad down, in case nature calls, but life has changed for the better.

Summary

- Child safety gates are useful for creating barriers that keep animals safe, whilst allowing them to become accustomed to each other
- Crate training should be taken slowly, with just short periods in the crate at first
- Gradually build up time in the crate, making sure it's always a rewarding place to be
- Don't use a crate as a time-out or place of punishment
- Only close the crate door for short periods at first, building up the time gradually

5 Who's walking who?

Kac

In my dog-owning fantasy daydreams I had always pictured having a dog who would run freely into the yard to play, and, when hearing my voice calling, would obediently return with a lilt in their step, and ears flapping in the breeze. Dreams are fun, aren't they?

When I picked up Talulah from her foster care mom, our conversations centered on Talulah's background, and I was emphatically cautioned to have a fence high enough to contain her. Talulah had recently been returned to a high-kill shelter by the owner(s) who had gotten her from the same shelter two-and-a-half years earlier, when her name was Bianca, and she was just a young puppy. She apparently lived with the same family until they decided to return her, and, from her semi-wild behavior – leaping, jumping and dashing off on her own – it appeared that she'd not had any real training. It was hard to say what her day-to-day life had been like, but when she was transported to a dog adoption event by the shelter folks, she got loose from the handler and dashed off into the Pasadena terrain of houses, byways, and highways.

Lisa Young, head conductor for the Rescue Train Organization, watched in horror, then chased after Bianca but could not find her anywhere. She returned several times over the next few days to try and locate what she believed would be a terrified, hungry, and possibly injured little dog.

Lisa had put out a dog APB, and local vets were made aware of the need to watch for her.

Three days later Lisa's phone rang. It was a Pasadena vet, who said that they had found the little dog she was seeking. Bianca had been hit by a car, sustained bruised organs, but was otherwise physically intact. The people who hit her said she dashed out into traffic; compassionate people that they were, they took her to a nearby vet and asked for help.

When little Bianca was due to be released Lisa went to the vet's office, paid the vet bills, and took her to a foster care volunteer mom, which is where and when I met her – post healing and calmed by the loving hands of Paula Cwickly.

As I listened to this tale of woe, I was shocked, grateful, and a little edgy about Talulah running amok again. Lisa reminded me to make sure I had high fences because she was a jumper. She also suggested I get a GPS collar with

Talulah ready to go for a walk ... her way.

an electronic signal, in case Talulah wandered off, and cautioned me to always keep her on the leash. I promised: actually, I had to sign a piece of paper, saying I would comply.

When I looked at Talulah, I didn't see a wild renegade, but a sweet-faced, perky little being with long, model legs, and a personality to die for. Under that sweet Doris Day demeanor, however, lay the spirit of a triathlon sprinter who wanted to run and run until she had circled the globe three times in the same day!

I learned about these tendencies the very first time we went for neighborhood walkies, when Talulah acted like she wanted to break the sound barrier. I was into strolling; she was into rocketing. The first walk we ever took I had to ice my arm from all the pulling and restraining that went on.

By the time we were halfway around the block my opinion of the little dear was somewhat changed. I was trying to juggle being her parent, her friend, and her playmate, but nothing was working. Talulah had a mind of her own and a will of steel.

I bought a can of something called Pet Corrector, which emitted the sound of rushing air, to try and distract her from her manic behavior. Aversive dog training, is what's it's called, and it worked pretty well for a while. But it was becoming more and more obvious that Bianca/Talulah had zero training, and may even have been tied or restrained, which meant that, to her, freedom was running away. At least that's what my pop psychology notions concluded.

Talulah was not being deliberately awkward: she simply wanted to explore the world at the speed and force of a Tsunami. I was beginning to think she was part-kangaroo! I knew I needed some help figuring out how to modify her behavior so that walks were enjoyable, and not a power-struggle. The pet store sold me a retractable leash that only served to give me a deep rope burn, and

Talulah's idea of a walk was to pull me along behind her!

was, as I later learned, the wrong device to use with an uncontrollable dog.

In desperation, I emailed Lisa: H-E-L-P! (I'm sure that, by now, Lisa thought I was about the lamest person alive, and going about problems the wrong way.)

Lisa

Recall (coming when called) is one of the most important things we can teach our dogs – in fact, it's *the* number one task on my list with every dog I've lived with and worked with, as it helps keep them safe, and can even save their lives.

Talulah was fortunate to escape serious injury when she headed off into the unknown, having given her handler the slip, but other dogs are not as lucky. Reliable recall also means that you can call your dog away when she's finding the attentions of another dog unwelcome (or is the perpetrator herself), and when she's making a beeline for something that you don't want her to have.

Most dogs, given the opportunity, will race off to explore the world, and, when you think about this, it's really not so surprising. Dogs have to live by our rules. They're confined in the home most of the time, and when they do go out for walks they're often restrained from roaming because they're on-leash. Life can be boring for them if we don't make sure to provide adequate mental, as well as physical, stimulation, and the outside world is an exciting place.

So how can you teach your dog to choose to return to you?

The key to this is finding the right motivation. Is your dog a food freak? If so, that's great! You can use extra-tasty food rewards every single time she comes when you call her name. Not interested in food? A game with a tug-toy or a ball should do the trick. Consider what floats your dog's boat the most, and then use that only as a reward for coming to you when called. This makes the reward special, it gives it high value, and your dog is more likely to want to race over to you if she understands that something good will happen when she does.

Teaching recall needs to begin at home, where there are fewer distractions. Call her to you from across the room, and reward her handsomely as soon as she complies. Once this is working well, play hide-and-seek around your home, graduating to calling her in from or out to the garden. Keep her on-leash while out, though, until you are sure that her recall is reliable.

Use a long training lead when you go out, and practise letting her move away to sniff around, then periodically call her back to get her reward. Wait until she comes to you every time, regardless of the presence of other dogs or lots of tempting smells, before you let her off-leash.

A common mistake made by many is to only clip on the leash at the end of the walk, which, of course, signals the end of fun-time for your dog. If this results in her being reluctant to comply, prevent this problem by calling her every so often, rewarding her, and clipping on her leash; then releasing her with a 'Go play!'cue. In this way, she'll never be sure when it's the end of the walk, so will be happier about returning to you.

On the subject of leashes, I don't ever recommend extending leashes. As Kac found, they can cause nasty injuries, and I've heard numerous accounts of people receiving deep flesh wounds when a dog has run around them. If another dog runs into, or gets caught in, the leash there's the likelihood of serious damage.

Another aspect of extending leashes is that they're always taut. Dogs will automatically move away from anything that causes pressure, so these leashes actually *teach* dogs to pull – the opposite of what you want!

Walking nicely

As Kac found, the Pet Corrector didn't prevent Talulah from pulling on-leash in her excitement at being out in the wider world, and most likely this device would actually have increased Talulah's desire to keep a safe distance between her and her new mom. Aversive methods such as Pet Correctors, rattling jars of stones, shouting, spraying with water or jerking the leash do the opposite of what we, as guardians, want – they generate fear and mistrust, and compel our dogs to want to avoid us instead of be close to us.

The relationship we want to build with our dogs is one of trust and mutual affection: once this has developed through using kind, force-free methods, and an understanding of dog behaviour, our dogs feel relaxed and comfortable around us, and, when given clear instructions, will choose to do as we ask.

If you want your dog to walk on-leash without pulling, a long, soft leash is best. You can wrap it around your hand to reel your dog in if there's an emergency, and you can keep it nice and loose, giving her plenty of freedom to explore, safe in the knowledge that she can't disappear over the horizon. These leads are great for teaching recall, and they also give your dog some much-needed freedom.

Your dog will be more inclined to walk alongside you if there's a reward in it for her. After all, you need to be her motivator and a source of pleasure. If she pulls on the leash, simply stand still and wait until she looks around and moves towards you, or turn in the opposite direction, and move into a trot as you call her in a happy, excited tone of voice so that she follows. Each time she comes close, give her a reward and plenty of praise, which will make her far more likely to stay near you. Keep doing this, and she'll soon want to trot along beside you, especially if you also give her the freedom to move away and follow her nose to snuffle up all those heavenly scents that are calling to her.

Other dogs

Some dogs love the company of canine friends, while others become anxious and scared if a strange dog comes over to say hello. Dogs have a distinct manners code, and a dog who hasn't learned how to interact politely can either cause fear, or prompt a reluctant new friend to tell her to back off by growling or snapping.

Social skills learning begins during infancy via interaction with littermates, and introduction to new stimuli, such as strange noises and attention from visiting humans, during the weeks before the puppies go to their new homes. The more positive experiences a puppy has, the better this is, as these help her social development, and accustom her to accept the presence of unfamiliar people and objects as she grows and develops. Many cases of fear-based dog aggression are the result of poor socialization during the first critical period of learning (up to the age of sixteen weeks). Negative experiences with other dogs, such as being bowled over or attacked, can lead to lifelong fear issues, so it's vital to protect puppies when they first begin meeting other dogs.

Canine body language is a complex and fascinating subject, and, as it's a dog's primary mode of communication, it's important for all dog guardians to take the time to learn at least the basics. You can then tell when your dog is happy, worried, afraid, angry or sad, and take steps to help her. After all, as guardians, it's our job to take care of our dogs, and protect them from harm; to be their champions.

In silent dog-speak, polite greetings involve moving in a curve, rather than dashing head-on towards another dog. Curving is a signal that shows a dog's intentions are friendly, whereas a direct approach, especially when made at speed, is intimidating, as it can mean that a dog has aggression in mind. Butt-sniffing is a polite way of greeting because this divulges a great deal of information to the sniffer, who usually also takes turns at being the sniffee, and it's done at a good, safe distance from the toothy weaponry at the head end.

When a dog is anxious about interacting, she may display a number of signals, even if she doesn't make a sound. Just some of these are –

- moving away
- turning the head sideways
- looking away
- lowering the body and tail
- cowering
- closing the mouth
- panting
- pinning back the ears
- showing the whites of the eyes (whale eye)
- lip-licking
- yawning
- displacement activities such as sitting, lying down, scratching, sniffing the ground, and cleaning the genital area

If you see any of these, it means it's time to remove your dog from the source of her discomfort.

A dog who wants to make friends will show this by having a loose, open mouth, and soft, bright eyes, and by approaching in a curving motion. Tail-wagging can be confusing to guardians, because angry, as well as happy, dogs wag their tails. A low, wagging tail indicates friendliness, whereas a high, stiff tail that moves in fast jerks means trouble.

If your dog looks comfortable and receptive when meeting another dog, let them greet each other, but please, if your dog appears anxious, immediately remove her from the situation, or ask the other dog's guardian to call their dog away.

I was involved in a very interesting discussion between several friends a while ago, about what approach to take when asking people to step away from reactive dogs, or when asking them to call away a dog who is giving unwanted attention. This is something I always have to discuss in depth with clients whose dogs are reactive to unknown people or other dogs. I had plenty of personal experience with this when my feral dog, Charlie, began going for walks because, at first, he was extremely fear-aggressive towards people and dogs we saw whilst out. He would twist around on his harness, lunge, bark, growl, howl, and show his teeth. Essentially, he would exhibit every behaviour in his repertoire that could possibly result in ensuring a safe space was maintained between him and what he viewed as a threat.

How to protect your dog from unwelcome attention

It can be uncomfortable, unnerving, and sometimes even downright scary to be forced into a position of having to protect our dogs. But we are our dog's guardians and champions, and it's our responsibility to keep them safe, as well as others in the environment, too.

These are the most common scenarios –

• A bouncy dog, or one whose intentions seem unfriendly, charges over to rudely sniff at or jump on your dog. If your dog is nervous or fear-aggressive, he's likely to react strongly to the intrusion. He may seek an escape route (which won't be available if he's on-leash), or may cringe, bark, growl, lunge or even attack, in order to create distance between him and the other dog. If he's off-leash, he may bolt, and fear will keep him away when you call him.

• The owner of an over-friendly or intrusive dog either takes no notice (it's amazing how many people let their dogs off-leash, and then are too busy texting or chatting to keep a check on where they are!), or sees your concern and calls out 'It's fine! He/she just wants to be friends!' Even if the other dog *is* friendly, your dog may have issues around other canines, and may take offence at having his safe space invaded.

• An unknown person comes over to greet your dog, and pays no attention to the 'back off' signals he's giving out. Retreating, leaning back, dipping his head, lowering his body, widening his eyes, flattening his ears, raising his hackles, or stepping forward and lunging, jumping up or barking, are just a few of the signals dogs give to ask for space. Hearing the cry 'I love dogs, and they like me' can make the heart of any owner of a fear-aggressive or very timid dog sink like a stone. It's a sure sign that the intruder is determined to pet your dog, come what may, and take the risk of being bitten. If the dog reacts in an adverse way, the intruder is shocked and blames you or your dog!

• Someone with a dog insists on introducing *their* dog to *your* dog, and you're aware this could end badly.

So, how can we be assertive on behalf of our dogs, whilst remaining polite and in control of the situation?

The first thing to do is take a deep breath. Stay calm. If you freak out, shout, or otherwise react strongly and loudly, this will only make your dog feel even more unsafe, and it's likely he'll panic and become even more reactive.

Step between your dog and the unwelcome visitor (canine or human) so that you're a human barrier. Stand still. Stand your ground. Then follow through appropriately, depending on which scene, A or B, is being played out –

A Keep your body between your dog and the visitor dog: you may have to do a little dance as both shift position – one to get closer, and the other to create distance. If you have a free hand, hold this up, palm out in the classic 'stop' position,

and firmly say 'Off' to the intruder. If the owner is close by, ask them to call away their dog, adding that your dog is scared and could react unpredictably. Usually, the owner will call off their dog and you can walk your dog away immediately.

If the owner is further away, wave your arm at him or her to gain their attention. Jump up and down a little if you have to, while dodging between the dogs. Most people will call away their dog at that point, even if they then give the 'But he was only being friendly' excuse. If so, you can explain that you're teaching your dog to cope with the presence of other dogs. Usually, people will then be understanding, rather than hostile.

B If the owner is otherwise occupied, and totally unaware that there's a problem, you need to catch his or her attention. If other people are around, ask them to signal to the owner. If you're alone, you'll need to raise your voice and call out 'Excuse me' (or 'Hey!' if they're still oblivious), then ask them to remove their dog. If you hear 'He's just being friendly' when the dog clearly has other intentions, you can calmly point out that your dog is scared and not friendly; name the body language that shows their dog has a problem with your dog, and say that the situation is close to getting out of hand. Usually people will move their dog away in this situation, though they may grumble a bit. If they want to blame your dog, don't worry about it. Walk away. At least you've kept your dog (and possibly theirs) safe, and an argument between owners wouldn't be good for either animal.

• Uninvited attention from dog-lovers can be a real concern when your dog is anxious. It's a rude intrusion on his personal space, and is as unpleasant and unnerving for him as it would be for you if a stranger suddenly walked over and grabbed you by the shoulders. If someone targets you with that 'I love dogs, and want to stroke him' gleam in their eye, step swiftly between the intruder and your dog. Hold up your hand in the 'stop' position, if necessary. Calmly and pleasantly explain that your dog is very nervous of strangers, and will react if they come too close. Usually, this works.

However, I *have* experienced situations where someone has kept on coming, asserting that all dogs love them. Stand your ground, and repeat that your dog is reactive when approached too closely. You'll know the limit of your dog's 'safe zone' (the area of space around him that he needs in order to feel comfortable), so ask the dog-lover to remain outside this zone, and be willing to chat for a minute about your dog, their dog, or any other dog! The longer the intruder is there, without touching or looming over him, the more likely it is that your dog will learn to become more comfortable around new people.

Explaining this often makes other people feel useful and helpful – you can say that they're helping you with his training – and often they leave feeling good about themselves, and rather well-disposed towards you!

• I've often had it happen that people with dogs want their dogs to become friends with my dogs. Skye, my Lurcher, is very open to this, but Charlie needed careful introduction at a distance for several weeks before he began to view unknown dogs as potential friends, after which, he was eager to say hello. You can't force dogs to become friends, any more than you can insist your child likes another child. Sometimes, there's a bond at first sight; other friendships take time to blossom – or never develop at all. The key point is to never, ever force interaction between dogs. It puts them on the spot, makes them feel vulnerable (especially if they're on-leash), and can lead to conflict.

If someone is insistent about an

introduction and you don't feel it's wise, calmly say that it isn't yet the right time, but you'd be happy to set this up later, when your dog is ready. If you feel okay about it, then set up a plan of action with the other owner. Start off with the dogs walking parallel, on-leash and at a distance, with the humans in the middle, shielding the dogs. Move closer when both dogs seem comfortable, then change position so that the dogs are walking parallel in the centre, with you on the outside. Let the most reactive dog follow the other dog for a while, to catch drifts of his scent. If both dogs seem relaxed, let them greet each other. Use the Three Second Rule, which involves allowing no more than a three-second greeting before moving the dogs apart. If both dogs seem happy, bring them together for another three seconds.

Keep repeating this until the dogs make it clear that they want to play. Observe their body language and move them apart if you notice any signs of tension: tense posture, eyeballing, ears right back, tail raised, and hackles up are just a few of the signals dogs give. If they seem to be enjoying getting to know each other, that's great. If not, then explain it's not going to work out, and walk away.

The Yellow Dog project in the UK and USA is raising awareness that some dogs need space, and owners can use a yellow ribbon or bandana as a marker that their dogs are reactive and should be steered clear of. This is a useful way of protecting dogs who are anxious, nervous, fear-aggressive, unwell or in heat, and I hope it will catch on around the world.

Kac

Our walks became more manageable. I didn't have to put gel muscle rub on my arm when I got home, and we now walk as a team. I give Talulah freedom to explore, and she minds me when I ask her to 'wait.' We're working on encounters with other dogs. Talulah barks at them, but I don't think it's aggression: she's friendly and excited and wants to play.

Six months after putting Lisa's tips into action I was taking Talulah out to the car on her leash, my arms loaded with purses, bags and 'stuff.' Talulah was walking perfectly well with me, but I didn't see a step, tripped, and fell down on my knee, while losing hold of the leash. "Ouch and dammit," were the words I heard myself emit.

My first thought was of Talulah. Oh no! I had dropped the leash, and fully expected her to bolt into the forest and up the neighbor's hill. But she didn't. Instead, she turned, saw me prone on the bricks, and came over to check on me. She didn't bolt; she didn't race off: she simply waited for me to gather my fractured person and headed right back into the house with me to search for an ice pack. There was no drama, only sweet companionship from my dear Talulah, the now-tamed kangaroo.

Summary

- Reliable recall is one of the most important things we can teach our dogs
- It's best to begin teaching recall lessons at home where there are fewer distractions
- Dogs learn best when we motivate and reward them
- Aversive methods erode trust, and should not be used
- A long, loose leash is a great training aid for dogs who pull on-leash
- Social skills need to be learned early, and continued
- Understanding your dog's body language helps you to know when she is happy, anxious or scared
- There are simple steps you can take to protect your dog from unwelcome attention

6 Talulah goes to school

Kac

Talulah is a multi-faceted dog. She has the spirit of a champion, the legs of a gazelle, the heart of Mother Theresa, the curiosity of ten cats combined, and the courage of a Viking. Try keeping a lid on that! Whatever she does she throws herself 310% into it.

After several less than successful disciplinary attempts at home, we decided the best thing to do was enroll Talulah in training classes with a professional dog expert. We looked for positive dog training classes and found one locally. I had read an article about abusive dog training practices, about which some of the most respected canine behaviorists and trainers are concerned, and the more I checked into these training techniques the more I wanted to look thoroughly to find the opposite. A local pet store offered positive training classes for puppies and canines of all sorts on several levels.

My fantasy was that I would drop off Talulah for class, go across the street for a nice cappuccino, and come back in an hour to collect my perfected puppy. I was anticipating her having the equivalent of Emily Post pooch manners the instant I arrived.

I was informed, during registration, that her human parents were also expected to attend the classes, as they too, required some training. Drat! This messed up my plans, but I was willing to do it if it would help Talulah. I signed on the dotted line, and Talulah became a schoolgirl.

During the first class we were instructed to sit across the room from each other so as not to distract her from the dog trainer, but Talulah had other ideas, and spent the whole hour running from one side of the room to the other, jumping up on our laps and ignoring the trainer completely. She could not have given a rat's patootie for his commands, his treats ... or his goatee.

At the end of the session his indirect understatement, via a deep sigh was, 'I can see I have my work cut out for me.' We weren't sure if he meant Talulah or us.

By lesson two Talulah was hip to the jive. She knew the way to the training room in the back of the store, and pranced her way past the fish tanks and parrot cages all the way to the back, to greet her trainer like they were long-lost friends. It was a nice relief.

Training began with a simple 'sit' cue, followed by a treat reward. (The

Talulah meets her trainer the first day of class.

Talulah begins her training and learns how to sit on cue. Treats are involved.

secret to the success of the second meeting may have been as simple as the trainer upgrading the quality of the treat – one that Talulah preferred over the other.)

Talulah mastered 'sit' in short order, so her trainer went right on to teaching her 'down' (paws stretched out on the floor and butt down). Talulah followed this with reasonable ease. We were brought into the session and asked to repeat the cues. 'Sit' was easy but 'down' was a challenge. We were shown how to hold the treat in a way that encourages the dog's nose to follow it to the ground, without giving the treat until the deed is accomplished. By the end of the session, we were managing both cues and Talulah was full of treats. We got it!

The subsequent five training sessions consisted of the trainer training us and our dog in the art of the cue: execution, reward, and patience for the number of repetitions it might require for Talulah to connect the sequence of actions and behavior associated with the reward/

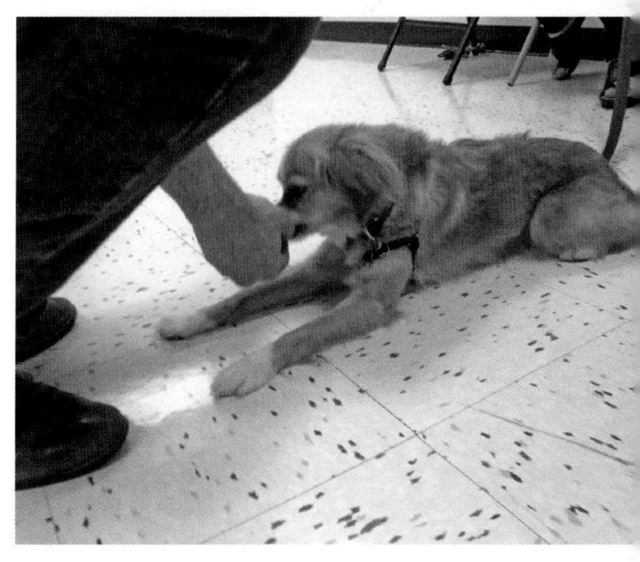

Talulah learns the 'down' cue using treats as a motivator.

treat payoff. I cannot say it was the most fun I've ever had, but seeing the progress Talulah made definitely warmed my heart.

By the time we reached week seven she had mastered 'sit,' 'down,' 'wait,' and 'okay.' 'Okay' was for waiting in place until she was released, and we started with twenty seconds and worked our way up to the two minutes that Guide dogs are taught.

Having tasted the allure of success, we eagerly signed up for a second round of classes for Talulah. Once again, I pictured her twirling in tutus across well-known stages, and one day even auditioning for Simon Cowell.

In the next set of classes we learned more about 'wait,' near sliding glass or automatic doors; we learned to stay on curbs, we learned how to stay 'down' and not jump up, and we learned to get down (furniture or places of danger) when the 'off' cue was given. We also learned more about interfacing with other dogs that we meet and greet on the street.

Talulah's grasp of action-associated behavior and rewards improved each week, and two things happened for me. I took the short dog guardian's course – the Dog's BFF [Best Friend Forever] Award Course – that Lisa and Dale McLelland wrote for the ISCP (International School for Canine Psychology & Behaviour), which gave me a great introduction to the way a dog thinks, as well as insight into my role as gentle guardian – rather than drill sergeant – of my ward. Secondly, after becoming a BFF to Talulah, with a certificate to prove it, I learned that, among other things, I didn't really have to SHOUT a command. Her hearing was perfectly good – three times better than mine, actually – and she responded equally well to a soft vocal cue as she did to a staccato bellowing so loud that the entire neighborhood was privy to it. (She probably thought I was reliving a past life as a stern military commander, barking hostile orders to a wide-eyed, and partially deaf, rookie.) I changed my evil ways.

Talulah received training and so did we, and this may have been the most beneficial thing we ever did, after adopting her. The training, encouraged by Lisa, gave her self-confidence, and inspired her to become the best she could be. Talulah is positively thrilled to cooperate, and sticks out her little chest in pride every time she does something to merit a treat.

She was such a good student, and a charming coed, that even the trainer fell in love with her. We made her a custom graduation cap, and had a little ceremony on her last day – the day she graduated from Puppy Training Level Two. We celebrated with champagne; Talulah got a chicken patty. We still have some more work to do to address the 'pulling on the lead' tendency, but, overall, it was a rip-roaring success.

Lisa

Training classes can be great for dogs – providing you attend the right class. If you're considering taking your dog to school, thoroughly check out the classes in your area first of all. Not all trainers have a qualification, and you'll want the best for your dog. It can help to have a checklist of questions.

- Does the trainer have a qualification from a reputable education provider, and is he or she a member of a professional organization? A qualification and organization membership shows that the trainer has studied, and has agreed to observe a member organization's code of ethics, which should give you confidence in the trainer's abilities

- Does the trainer use only force-free, science-based methods? Beware of 'balanced' trainers, who employ a mixture of aversive/punishment/discipline and force-free methods. Once you've begun classes, it can be hard to speak out if you feel the trainer is using intimidation

Talulah, and with her trainer, Chris, on Graduation Day, May 2014.

techniques, so let your feet lead you out of there if you feel your dog is being subjected to unkind treatment

- How big are the classes? A large class means that less time can be spent on individual dogs and guardians, and interactions between excited dogs can easily get out of control, and escalate to fighting. A small class, with plenty of space between the dogs, is more manageable for the trainer and guardian, and less intimidating for the dogs

- Do the dogs in the class seem to be enjoying themselves, or are some clearly anxious or nervous? If any of the dogs are staying close to their guardians, cringing, trying to hide, and displaying body language that shouts out 'stress,' this should flag a warning

- Will the trainer allow you to come and watch a training session without your dog, so that you can get a feel for how they work? If not, it's best to look for another class

There's more awareness of the reality of our relationships with our dogs now that the scientific research into dog behaviour and training is being widely shared among dog guardians. Our dogs aren't trying to be the 'alpha' or to 'dominate' us. They have no desire to rule the roost (though, of course, given plenty of license to do as they please, they will make the most of whatever advantages life – in the form of you – offer).

Dogs need to learn good manners and doggy etiquette, and (like children) they need some rules and boundaries so they don't run amok and create mayhem. These can be taught in such

a way that your dog finds learning fun, and classes are great for learning in a social environment. Of course, you can do it yourself at home, too, and if you do attend classes you'll need to keep up the training in-between sessions: if you don't do your homework, your dog will soon forget her lessons.

'Sit' and 'wait' are the next most important lessons after a reliable recall, and 'come' is linked to recall. If your dog jumps up at visitors, 'sit' provides an alternative activity that keeps her paws on the ground and teaches polite greeting manners. Sitting while waiting for a meal to be served, or for having her harness or leash clipped on, helps her learn to be calm at times she finds exciting. 'Wait' is useful for preventing her from dashing off when she sees or smells something interesting.

Sit

The easiest way to teach 'sit' is to call your dog to you, and hold a treat just above her nose. As her face turns upward to seek out and take the treat, her butt will tend to lower to the floor. The moment this happens, give the treat and fulsome praise.

After she's done this a few times, begin to pair the action with the word, and say "sit" just as her butt touches the ground. Make sure you give the treat while she's still sitting, because if she jumps up to receive it she'll associate the 'sit' cue with getting up. Timing is everything! If you decide to use a clicker, you can mark the moment of butt impact with a click, and then treat instantly. The sound of the click tells her that she's done the right thing. Celebrate every time she gets it right!

Wait

This cue needs to start off small. First, ask her to 'sit' and reward her for doing so. Hold out your hand, palm out in the traffic cop 'stop' signal, and take a step or two back as you say "wait," then move forward to give a treat before she gets up to follow you. As you practice this, you can gradually increase the distance from her, but please bear in mind that it's important not to expect too much from your dog by trying to rush the training. Slow and steady is most effective.

Okay

You can use 'okay' to release your dog from a sit or stay position.

Come

'Come' is useful for getting her attention, and for learning recall. Make yourself inviting. Lean forward and pat your thighs, or crouch down and hold out your arms as you say "come" in a welcoming, happy tone of voice.

Down

'Down' is very helpful for trips to the vet for examinations, or to teach her to lie down to be groomed. You can also use 'off' if your dog's bouncing around or jumping up. To teach 'down,' hold a treat between your fingers in such a way that it isn't easily accessible. Gain her attention, let her see you have something tasty, and lower your hand with the treat in to the ground. She'll lower her body to get to the treat, but you may have to move your hand along the ground until she achieves a full down position. Once her body is down, say the word as you give the treat.

If she doesn't figure out what you're asking of her, try another method. Sit on the floor with your knees bent, so that there's space between your legs and the floor. Show your dog the treat, and tempt her under your raised legs so that she has to flatten her body to go under them to get the treat. As soon as she goes into the 'down' position, say the word and treat her. After a few sessions she will have figured it out, and you can then stand while you give her this cue.

As Kac discovered with Talulah,

there's no need to raise your voice. In fact, a quiet voice is best because your dog's more likely to focus on you. Think about how *you* respond to someone's tone of voice. If they're shouting at you, your first impulse will probably be to move away, because a loud, harsh tone creates tension. A quiet, calm tone of voice is more likely to get your attention and persuade you to listen.

We're teaching our dogs constantly, even when we're not aware of doing so. Dogs watch us very closely. Their survival and wellbeing depends on us, so they watch our body language, listen to our tone of voice, and look for cues to what we're asking of them.

Dogs are the only animals, other than humans, who use what is called 'left gaze bias.' When we're with another person, our gaze naturally strays to the left when we're face-to-face, to focus on that person's right eye. It's considered that the right side of the face reveals more about our emotional state, and experiments at the University of Lincoln in the UK, carried

The Graduate was now sufficiently well trained to safely accompany us on a scenic boat excursion. She went right to the helm and claimed the Captain's seat.

out by scientists Professor Daniel Mills, Dr Kun Guo, Dr Kerstin Meints and their team, have revealed that dogs also use left gaze bias to decipher our moods and feelings. They even do this when shown upside-down photos of people, but they don't do it with other dogs, interestingly. This indicates how important it is to dogs to be able to understand our emotional states. It's likely that this behaviour evolved as a way to keep themselves safe (a frowning face is a signal to back off), but it also shows how important we are to dogs, and how much effort they put into understanding us.

Clear communication is vital to all relationships. After all, if we can't find ways to 'read' each other, we don't have a foundation on which to build. We tend to expect a great deal from our dogs, and it's easy to forget that they're trying just as hard (and often harder) to communicate with us as we are with them – and that we speak very different languages! The Dog's BFF Award course that Kac took was designed by Dale McLelland and me to open up the dialogue between dog guardians and their dogs, and give guidance on responsible guardianship, and the course has a strong focus on how to interpret canine body language.

Talulah's graduation from training class was a joy. I know how dedicated Kac and Marlene are to her, and it's exciting to know that a lot more developments will blossom in their relationship – including walking on a loose leash!

Summary

- Check out training classes thoroughly before you enroll in one
- Our dogs have no desire to dominate us – this is a myth
- Dogs, like children, need fair rules and boundaries
- Training can be fun for both you and your dog
- Dogs watch us closely and are always learning from us
- Clear communication is vital

7 Big sticks, garden hoses ... and men in baseball caps

Kac

We didn't receive a lot of detail about Talulah's home life when she was Bianca, but we concluded that there must have been some kind of abuse because there were certain things she reacted to by cowering, running away, sticking her plume-like tail between her legs, and looking back over her shoulder with abject fear in her eyes. It was an alarming discovery.

The first time I took a brush out of the drawer to comb her, she darted off into the corner and looked at me as if I was going to hurt her. I was shocked, and didn't know what to do. I tried walking up to her, telling her softly "It's okay, I'm not going to hurt you." But past conditioning was so strong that she simply shook with fear. I put away the brush.

In the training sessions I had learned a little bit about the power of treats, so I trundled out to the kitchen, pocketed some bribes, and slowly took the brush to her again. This time I used it on myself first, showed her it was a good thing, and rewarded her with treats.

Talulah eyed the brush with a good deal of suspicion, but I remembered that presenting her with her favorite treats was the key to success. She gradually let me comb her, and eventually got used to the brush. However, even now, when I first take the brush out of the drawer she is startled every time. Although she doesn't run and hide anymore, it takes her a moment to remember that this is an okay thing.

It had become obvious that sticks or long-handled anythings terrified her. Whenever I took a broom to the deck to give it a sweep, she ran, hid, and retreated in fear. We tried to ease this emotion in her by one of us giving her treats while the other moved the broom. After months of treat therapy, she's better, but not totally cured of the old associations.

The day I brought out the garden hose to water my potted plants on the back deck, you'd have thought I had invited a cobra to dinner. The second I uncoiled the hose and water came out of the spray nozzle, Talulah practically jumped out of her dog suit. Of course, I shut it off immediately, and went over to reason with her. I simplified ridiculously, "This is just water, Talulah, nothing to be afraid of." She stared at me with a trembling upper lip. Oh dear. That time I decided to put her inside and let her watch from the window. Maybe she would see that there was nothing harmful happening. No such luck. The identical reaction appeared the next time I tried watering with Talulah out on the deck.

The only idea I could come up with was to slowly and surely try to show her that I was not going to spray *her* with water, only the plants, and that she was safe. (I have to admit that, by this time, I was finding it very hard to forgive those who had terrified her so, whoever they were. I don't know for sure, but I have a feeling that Talulah was an object of entertainment for some unthinking people, who had no idea how to treat a dog, and enjoyed tormenting this sweet little being. I'm willing to be wrong ...)

Little-by-little Talulah has learned to be in the same area as the garden hose, and has regained her trust that I won't suddenly turn it on her.

Another phobia that astounded us was her fear of men in baseball caps. Her reaction is vehemently negative; she barks like a crazy canine, and bends her chest halfway to the ground, as if she's digging in for battle. If I can get to the man wearing the baseball cap and fill his hands with treats, we have a chance at success. If not, I am compelled to ask him to remove the baseball cap. She generally calms down a bit, then, but, without treats, she has no trust for the man involved. Men in ties are fine. Men in t-shirts are fine. Men in suits work, too. But put a man in a baseball cap in her sightline and she turns into Cerberus at the gates of hell.

It has taken me a while to get used to having treats in my pocket whenever we go places, but it's a must if I want Talulah to feel safe, even when her fear-triggers rear their ugly heads. Some trainers have special pockets or aprons they wear to carry treats with them at all times when they are training dogs. I haven't found an accessory that suits my taste as yet, but I'm looking for one. Maybe a treat dispenser like the coin changers that bus drivers wear would do the trick. I'd need one with some bling, though. I'm all for treats, even without an accessory. They seem to help change behavior, keep the peace, and provide comfort, even in the face of negative memories.

Lisa

There are a number of reasons why some dogs are afraid of specific things. Maltreatment is one, and the poor dog subsequently carries such a strong association between the fearful incident, such as being treated roughly during grooming or targeted with a garden hose, that even the sight of these objects cause the previous terror to resurface. However, although it's natural to assume that the dog has suffered through mistreatment, cruelty isn't always the cause. It could be due to genetics, negative early experiences, or lack of early experiences.

Genetics is a major cause of fearfulness, especially if your dog is scared of a number of objects. It's a sad fact that a fearful mother breeds fearful puppies, because the floods of stress hormones she experiences through pregnancy are transmitted to the pups. This is why, if you're buying a puppy, it's important to meet the mother, and it can help to meet the father, too. If the mother is scared, nervous, anxious or aggressive, there's a strong likelihood that her puppies will grow up the same, however kindly they are treated.

Another cause of fear is a lack of socialization and closely monitored exposure to a wide range of experiences during the first formative weeks after birth. Dogs go through critical periods in which they become more fearful, so it's vital to doubly ensure that they have good experiences during these periods.

Fear and stages of life

The first twelve weeks of life is the critical socializing period, during which a puppy should have a rich variety of carefully planned positive experiences with people and objects. As most puppies don't leave the litter until around eight weeks old, this means that responsibility for introducing new experiences and stimuli, so that the pups cope more easily with the transition to a new home, falls to the breeder. It's important that the puppy isn't overwhelmed, or subjected to what behaviourists call 'flooding,' which is prolonged or repeated exposure to something she is afraid of. This will set up an automatic fear response in the future that will necessitate a great deal of work to overcome.

My friends who run rescues sometimes take in pregnant dogs and small puppies who have been heartlessly abandoned, and we've had a lot of discussion about how to help some pups who were gestated in a bad environment, and went on to exhibit behaviour issues after they had reached maturity, despite receiving the best of care in their foster and adoptive homes. The impact of the mother's mental and emotional state simply cannot be underestimated.

Happily, though, the positive effects of environmental enrichment – even on puppies with an unknown background – can set up a dog for life, generating confidence in a variety of situations and environments. My rescue friends who are caring for pups turn a room into an obstacle course, filled with items that make squeaky, scrunchy, rustling noses; that roll and move as the pups negotiate them. The pups have a wonderful time playing together, and all these unfamiliar objects just become part of the game. New people and children who are kind and gentle, and who handle the puppies appropriately, are introduced to them, and introductions to other animals are made under careful monitoring to ensure the experience is pleasant for everyone. Household objects such as the vacuum cleaner and washing machine are used, with lots of rewards given to create positive associations, and reduce the likelihood of fear issues later on.

The second critical period occurs during adolescence – the 'teen period' – when pups are hard-wired to seek independence as hormones kick in. This can start at around six months of age, and continues until maturity. In small breeds, adolescence passes fairly quickly, and the dogs are likely to move through it by the time they're eighteen months old. In large breeds it can continue up until the age of three years, and, sadly, it's at this stage in life when the most dogs are relinquished to shelters because of unwanted behaviours.

Training that has been successful up to that point may need to be set in place all over again. Behaviour can change, and there may be an increase in fear responses – and no wonder, if you consider how hard it is to focus when hormones are raging. If you have teenage children, you'll find it easy to understand why your adolescent dog suddenly becomes antsy, has periods of ignoring you, seems to suffer from selective deafness, and finds it much harder to cope in stressful situations without going into meltdown. Time, patience, and lots of positive reinforcement are key to encouraging the behaviours you want to see, and this stage *will* pass.

Skye, my Lurcher, is a large breed dog: a mix of Deerhound, Greyhound, and Saluki. He reached adolescence at six months, and my happy-go-lucky pup suddenly hurtled into forgetting what he'd learned, ignoring me when I called him, and exhibiting the hallmarks of teenage defiance. Because I understood what was happening in his body, I simply enjoyed the ride, kept up with his training lessons, and often had a darned good laugh (quietly, in another room) at his antics.

The turning point came when I was working on a book, and Skye decided he wanted me to play with him. As always, when it wasn't convenient, I said "Later, Skye," and carried on working. At that time, his favourite of several dog beds was a large bean-bed beside my desk. Skye stood and stared pointedly at me, trying to get my attention. I continued typing. Out of the corner of my eye I saw him move towards his bean-bed, look at me again, and then deliberately cock his leg and pee on it.

My response wasn't quite what he expected. I calmly stood up, moved the bean-bed into the garden shed without saying a word, and went back to work. Skye stood still in the now-empty space, looking hilariously surprised, then settled down on another bed and fell asleep. Later, after finishing the chapter I was working on, we had a game. It was Skye's final act of teenage attention-seeking, and I still smile about it, years later.

Old age is the third stage of life where fear issues can surface or intensify. Age-related health issues, pain from creaky joints, a reduction in vision or hearing, and the onset of canine cognitive dysfunction (doggy dementia) all bring challenges that are tough on our dogs. Just imagine how confusing it must be to lose track of the world through your senses, especially as dogs' senses are so much keener than ours. Our old guys need our total support and understanding during their twilight time.

It can help to keep the home environment unchanged, to avoid moving furniture, and to ensure there's easy access to water and food bowls, and beds. An obstacle course of shopping bags or toys on the floor can be scary for an elderly dog. Speaking your dog's name before touching him will alert him to your presence so that he doesn't jump at unexpected contact. Protecting him from the attention of other dogs outside the home will help him feel safe, and taking walks in a familiar area can help him feel more secure.

Like Talulah, many dogs are nervous of men (one theory behind this is that their voices are louder and deeper than those of women), of baseball caps or hats, and some dogs find uniforms and umbrellas scary. Usually, this is because these objects are unfamiliar, although, in some cases, a dog may have been scared or harmed by someone wearing the attire the dog now reacts to. Fancy dress outfits are an issue for a lot of dogs, because the familiar human suddenly becomes a stranger who smells the same but looks very different. One dog I worked with was terrified of airplane trails in the sky. Several more were scared of hot air balloons, and we get a lot of those floating past where I live. Fear of thunderstorms is common, and this isn't just because of the noise and flashes, either. Every hair in your dog's coat culminates in a nerve-ending, and the static electricity in the air before and during a storm can be horribly uncomfortable for them. Some causes of fear may seem weird or mysterious to us, but it's vital that we address them and help our dogs to cope.

Our human babies go through fear periods where they scream at the sight of a stranger, and try to hide behind us. We treat our children with understanding and compassion when this occurs, and it's only fair (and makes perfect sense) that we do the same for our dogs. We're their guardians and protectors, and we can gently guide them towards overcoming their fears when we use positive, force-free methods.

The chemistry of fear

Fear has a powerful effect on the brains and bodies of both us and our dogs. The source of fear (a sudden noise, for example) triggers an immediate response from the amygdala, a small structure in the limbic system of the brain that deals with powerful emotions, especially fear. The amygdala sends out an alarm response to the hypothalamus, which prompts the release of cortisol, and alerts the body to respond by running away or attacking (the flight or fight response). The heart races, blood pressure rises, and the muscles tense, ready for action. Your dog's rational brain (and our own) has no control over this; it's an instinctive, autonomic process.

It can take up to two weeks for stress hormones to subside, so, if your dog has had a fright or shock, it's important to keep her calm afterwards and avoid stressful situations so that she can relax again. Repeated exposure to the source of fear (behaviourists call this 'flooding') means that the stress hormones are continually building in her body, which can lead to all manner of physical health issues, as well as increased nervousness and anxiety.

Dispensing the goodies

Kac concluded above by mentioning treat dispensers to have handy at all times. These have been my article of choice throughout years of working with dogs. You can buy treat bags to put in your pocket or attach to a belt. I also have a 'bum bag:' a bag on a belt that clips around my waist, with plenty of room for treats, keys, poop bags and my phone. If you want to bling it up, you could get a small, pretty shoulder bag to wear and fill that with goodies. Be creative and make your own dispenser, or check out some of the many that you can find online and in pet stores.

Summary

- There are many causes of fear issues
- Genetics, early experiences, and lack of experiences are important factors
- Dogs go through critical periods where they may be more fearful
- Fear has a powerful effect on the brain and body

8 "I love Dr Suzy!" says Talulah Lake

Kac

It's always a blessing when local vets offer a free, first-time visit for all rescue animals, because, since there are six cats already living in our home, all complimentary services are greatly appreciated.

Talulah had a thorough veterinary work up, shots and health check before we adopted her. Her immunizations were up to date, and she was recovering from the internal bruising caused by the car accident. There was no real rush for a check-up, but getting her registered with our local vet was high on the agenda in case of an emergency.

Dr Suzy (Van Beurden) of Cambria Animal Medical Services in town is a fresh-faced, exuberant, and knowledgeable young vet. She tended our cats' health, and we enjoyed a wonderful rapport with her. On the day of Talulah's first appointment at the vet, I took her to see Dr Suzy. By nature, Talulah is a vivacious little creature, and, when she saw the scale in the doctor's office, she must have mistaken it for a trampoline, as she jumped up and down on it, making the reading bounce around like a carnival ride. Treats came into play, and she settled long enough for the receptionist to get her weight.

Talulah treated the waiting room like a gymnasium, and bounded around as if she owned the place: her own private playground. Oh dear. I was a little embarrassed as she bolted from cushion to cushion and armrest to armrest. The receptionist suggested that it might be the smells of other dogs which particularly excited her. She was being kind, but I desperately wanted Talulah to behave and calm down.

Talulah, however, was intent on sniffing the air, whiffing the cushions, and generally acting like a veritable canine detective. Unlike our cats, she appeared to have no fear of the vet's office, even though the air was full of medical odors such as antiseptic and peroxide.

All smiles, Dr Suzy came out to greet Talulah. She bent down to her level, offered her hand for Talulah to sniff, and had a treat or two scuttled away in her lab coat pocket. Talulah took to her, and her assistant/husband, right away, trotting off to the exam room with them. From this point it was a sunny walk in the park.

Talulah's first visit to the vet was a roaring success. Even when a booster shot was administered under her skin, Talulah smiled. What a relief it was to find a great vet and a happy dog who was

Talulah is hesitant about approaching Dr Suzy, so Dr Suzy comes down to Talulah's level ...

... and coaxes Talulah toward her with treats ....

... which Talulah finally accepts ...

comfortable with her new doctor, *and* received a clean bill of health to boot. A great day, indeed!

Lisa

Visits to the veterinarian are a source of stress for many dogs, and Dr Suzy's approach to Talulah is a wonderful example of how easy these visits can be if they're handled in the right way.

Talulah was able to bounce around and explore when she arrived, embarrassing though her excited behaviour was for Kac (and truly, veterinary staff would prefer to see a happy pooch doing acrobats than a nervous one). Dr Suzy took care to decrease her height so that she was less intimidating to Talulah, she introduced herself respectfully, and she offered tasty treats that got Talulah feeling good about

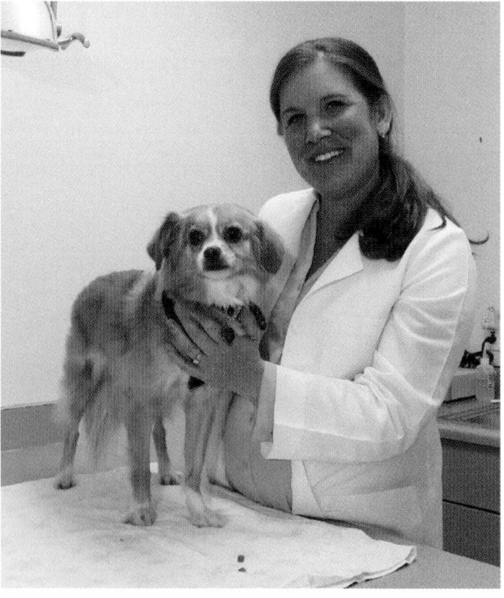

... and the pair become friends.

being around her. What a great way to help a dog feel comfortable! It's good to look at all situations from your dog's perspective, and ask 'How is this for you?' Imagine what it must be like to be taken to the vet surgery, and be immediately bombarded by the smells of chemicals, disinfectant, and the rank scent of fear left by other dogs. Scary. Add to this potent mix the appearance of a stranger who handles her in ways that she may not be accustomed to; who examines her mouth and body, and then perhaps gives an injection or medication. With such sensory overload and intrusion on your dog's personal space, it's no wonder that some dogs find the experience traumatic.

Good vet visits

Whether your dog is a puppy or newly-adopted adult, you can make vet visits into an enjoyable outing.

Firstly, choose your veterinarian carefully. Ask around for recommendations, then call the surgery you've decided on, and ask to take in your dog to briefly meet the staff. Most vets will be happy to allow this; after all, they want your dog to feel good about visits, too.

Take a pocketful of your dog's favourite treats with you, and go at a quiet time (the end of the surgery's working day, when the waiting room is likely to be empty, is the best time). Let your dog have a sniff around and introduce herself to the people working there, including the vet. Allow her to choose who she interacts with. You can help this along by giving some of her treats to everyone, so that she's rewarded for approaching. Some gentle strokes from your vet, paired with tasty food, will spark off a happy relationship quickly.

Keep the visit short, just five or ten minutes, and then take her home. It's likely that she'll be keen to go in next time, and will accept a health check without becoming unduly stressed.

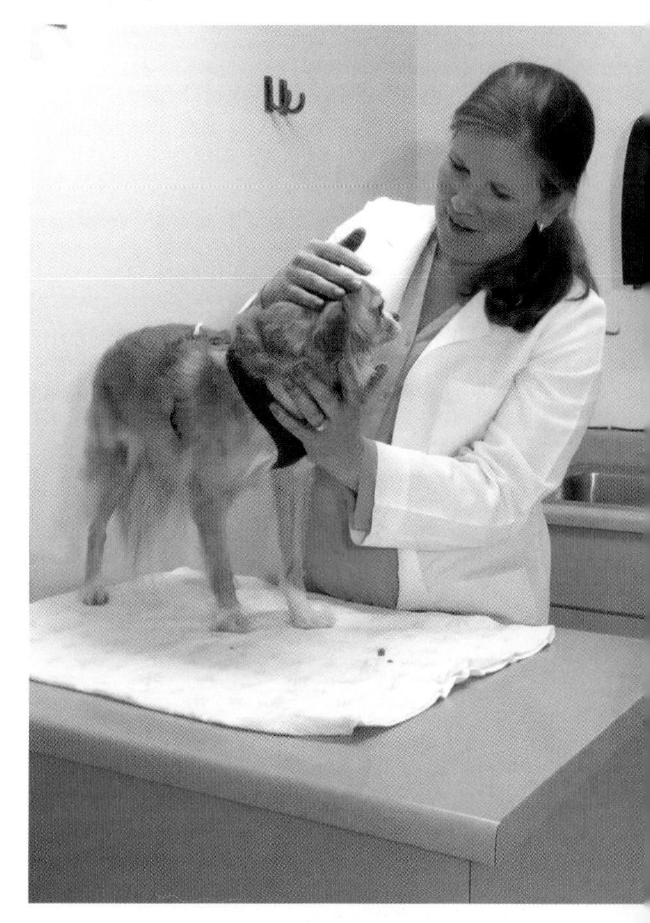

Talulah feels comfortable enough with Dr Suzy for her to perform a full examination.

Remember to take your stash of treats when she goes for her check-up and immunizations, and have these ready so that you can give her some during her examination. This will distract her from the unusual mode of attention she is receiving, and will further reinforce her recently-formed opinion that a visit to the vet is a highlight in her day.

Some dogs are less motivated by food and prefer a game with a favourite toy. If yours isn't particularly food-orientated, keep aside a toy that she really loves, and only bring this out during visits to the vet – bearing in mind

that you'll need her to stay still for her examination.

Police dogs are often trained to expect the reward of a tug game after they've done their work, and you can use this principle for your dog, too. Teach her to stand still at home, and run your hands over her, just as the veterinarian would. Pick up her paws, one at a time, to check her nails. Peek in her ears and mouth. Keep it short, and stop if she begins to look tense. You can gradually build up the handling as she learns to be comfortable and accept it. Offer the toy as soon as you've finished the exercise. She will then find it easier to remain still when the veterinarian is checking her over, and, of course, will then receive her reward as soon as the vet's completed the examination.

But she's scared of the vet!

If your dog is very nervous during trips to the vet, take her in several times for short introductions only before she has her first check-up. It's better to take your time and give her chance to understand that nothing terrible will happen to her than it is to try and rush things, and end up with a dog who quakes as soon as you approach the surgery entrance.

Some dogs may have had unpleasant experiences in the past due to rough handling, or being intimidated by another dog at the vet's, or through experiencing an uncomfortable or painful procedure. The key is to take things very slowly, and find ways to help your dog feel more comfortable. A dog who is frightened is unable to take food because the digestive system shuts down during the flood of adrenaline, cortisol and norepinephrine, the stress chemicals that spark off the emergency freeze-flight-fight reflex. If your body is signalling to you to be in fear for your life, the last thing you'd be thinking of, or interested in, is food, and it's no different with dogs!

Does your veterinarian do home visits? If so, this can help a lot because your dog will be more comfortable in her home environment. Is your dog happier in the car? Most vets are willing to do an examination from the trunk of a vehicle, providing there's enough space. Make sure you have plenty of tasty rewards handy so that the vet can offer these as soon as she arrives.

Does your dog have a favoured person, beside yourself? If so, you can ask that person to come with you so that your dog has additional support (and you do, too).

Complementary therapies can help dogs who are seriously frightened of veterinarians in general, though if you feel that it's just one vet that your dog is afraid of, it's sensible to see a different one. Bach Flower remedies, given the morning of a visit and afterwards, can be useful. Rescue Remedy helps relieve shock, fear and trauma. Rock Rose is useful for phobic states. Mimulus can help because it's given for 'known fears' that are specific and recognizable. You can add two drops of each remedy of choice to your dog's food and water on the morning of the visit. A drop of Rescue Remedy can be smoothed onto the dome of your dog's head when you arrive at the surgery, too.

When the vet becomes a friend

Skye, my dog, is immensely sociable, fortunately, because I have introduced him to a wide range of people, other dogs, and good experiences since he first came to me, aged eight weeks. He's met a lot of vets throughout his nine-plus years, and has had to undergo some very unpleasant procedures due to health issues, yet he still greets every vet he meets as if they're his favorite person.

Why is this? Quite simply because, right from the beginning, every vet has made a fuss of him, and I've made sure that there's always a bag full of his favourite treats in my pocket, ready to hand out just before the procedure begins – and during it. It really does pay to start early when creating a positive association with veterinarians!

Summary

- Veterinary clinics (called surgeries in the UK) can be stressful places for dogs
- Vet clinics/surgeries are filled with strange smells
- Giving your dog chance to explore can help her relax
- Ask for recommendations when seeking a veterinarian
- Keep the first visit brief and casual
- Take treats or a toy as a distraction and reward
- If your dog is very fearful, ask for a home visit
- Complementary therapies can help nervous dogs

9 I'm perfect just the way I am!

Kac

I love the garden-fresh smell of a newly-washed dog. It's as if they've been air-dried in a meadow, and fluffed by nature.

Talulah has a medium-length coat, and isn't one of those dogs who likes to lollygag and roll around in the mud. She does get hot from playing, however, and becomes a tad odiferant.

The local pet store had done a wonderful job with her positive training, so when I needed to find a place to get Talulah washed and groomed, I thought that trying out the pet salon there would be a good thing, so off we went.

Because Talulah was of the opinion she was perfect just the way she was (like I had told her so often), she did not feel the need for the bather/groomer's services, and was therefore extremely reluctant to be led away by said groomer. Not liking the idea of being separated from her mother, she looked back at me as she was being taken away to the bathing area, as if I had returned her to the shelter and was abandoning her forever. I felt guilty.

Three hours later I returned to collect her: a bouncy, flouncy, perky little lass whose bath and grooming had turned her into a total diva. She wanted to make sure the entire world saw her combed out ears and fluffy Pomeranian-esque tail. Whoa ... down, girl!

On a more serious note, prior to selecting a groomer or dog salon I had read a scary article about different types of dog groomer and their methods, which mentioned that injury and even death had resulted from dogs being chained to the grooming table. I resolved to make sure I selected a safe groomer, who genuinely cared for the dogs' welfare, and who wouldn't leave them where they might hurt themselves.

Lisa has more information about grooming safeguards.

Lisa

A dog's coat has natural oils, and, with short-haired dogs, a good brush or wipe down with a damp cloth is often enough to keep them clean – unless, of course, they've rolled in something stinky. All dogs need regular maintenance in the form of brushing and the occasional bath, and the longer and thicker your dog's coat is, the more grooming she will need. Brushing removes dead hair, reduces the risk of uncomfortable tangles forming, distributes the natural oils through the fur and skin, improves circulation ... and should be a

together through gentle stroking that includes touching her paws and various other parts of her body. Accustom her to the sound of the hairdryer so that this will be familiar if she is taken to a professional groomer.

The following tips are useful for dogs who are nervous about being brushed.

Talulah's spa day: Talulah and her groomer are good friends now.

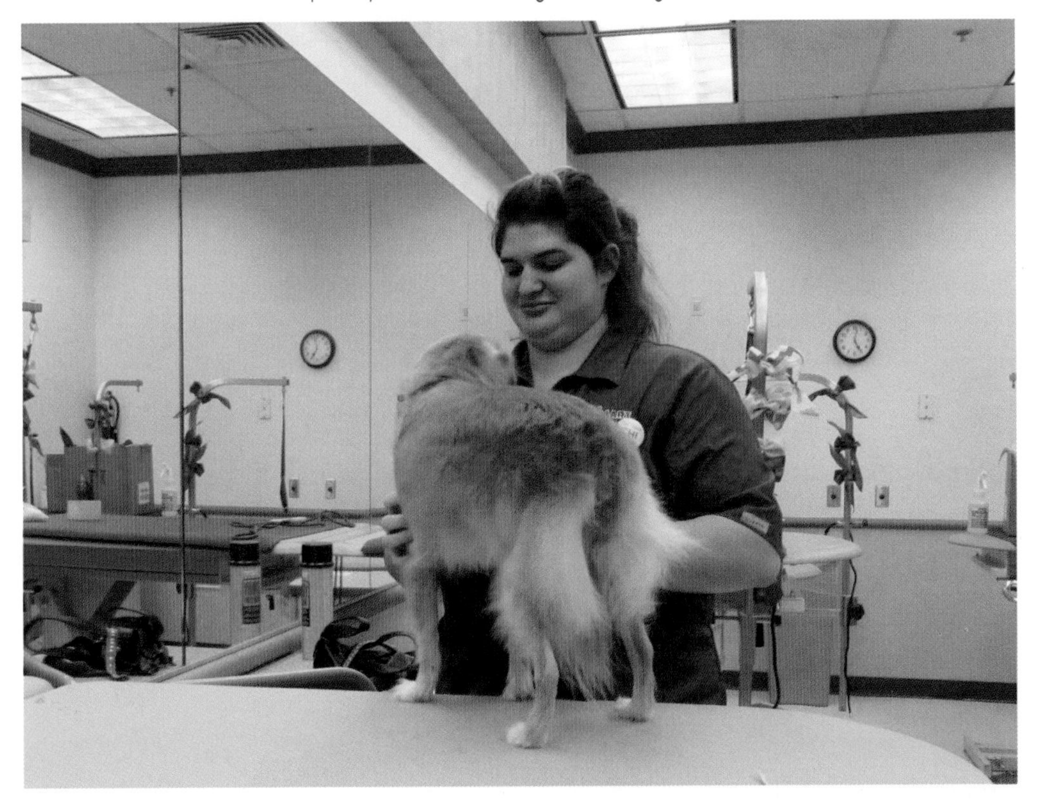

pleasant experience for your dog! Dogs with very thick or double coats, such as Newfoundlands and Akitas, need to be groomed frequently, as do Poodles and the Bichon Frise, whose curly fur can become matted.

Teach your dog to enjoy being groomed right from your first days

Introduction to grooming

A while ago, a group of my ISCP students came to visit for one of their practical study days, during which they work with dogs and their guardians under my supervision. As usual, I had found a family who needed some help with their dogs, and I took my students to meet their beautiful Golden Retriever and Labrador Retriever.

The family needed some guidance with teaching their grandchildren to be respectful towards the elderly dogs, which we discussed in depth.

Another question they asked was about grooming. The Golden Retriever hadn't been groomed for several years because she disliked the brush so much that she would snap at her guardians if they tried to brush her. Her rump, leg and tail feathers, especially, looked a little moth-eaten.

We got to work, and, to the family's amazement, their previously snappy dog hardly seemed to notice she was being groomed, and was actively enjoying the experience after just a few minutes. She even allowed her tangled rear end and leg feathers to be beautified, and at no time did she show any sign of finding the experience stressful. Since then she's been happy to be regularly brushed, and she looks much better for it!

Here's how this was achieved. Note it can help to have two people involved if a dog has previously been resistant to being groomed – one person at the head end, with lots of treats, and the other at the side with the grooming brush. You'll want to keep those rewards coming!

- Gather together lots of tasty treats and the grooming brush

- Call your dog to you, let her see the brush but don't touch her, and offer a treat as soon as she glances at the brush. Stroke her gently with your fingers on her favourite area (often, this is the chest, side or behind the ears)

- If she seems relaxed after you've given her a few treats, gently touch the brush against her side or chest for just a second, and give her a treat at the same time. Avoid touching the head area, and don't go as far as using the brush

- Then brush her side for just a moment, treating and praising her

- If she's comfortable with this, extend the brush strokes, keeping them light so there's no tugging

- Avoid leaning over her to reach the other side. This can make dogs nervous, so it's best to move right around her so that you're sitting or standing by the area you are brushing

- Keep the first session short, and try to avoid tangled areas this time, unless she seems to be actively welcoming the grooming process. Gradually extend the time spent grooming, and work carefully on trickier areas as she becomes accustomed to the brush, and keeps her focus on the rewards

- If a dog is happy to have most areas groomed but is resistant to one area, such as the lower back, consider what the reasons may be for this. Older dogs, especially, may dislike having areas groomed where there are arthritic changes, so if your dog seems uncomfortable, please take her to be vet-checked to rule out a medical issue.

Choosing a groomer

It's vital that you choose the right groomer. Ask around for recommendations; take your dog to visit before booking an appointment so that you can see how the groomer interacts with her, and, if you feel at all unsure, walk away and check out other groomers. It's better to travel further and feel confident that your dog's being well cared for than to take the first available option, only to then worry about your dog's wellbeing and safety.

Most groomers are dedicated to their work, and have wonderful relationships with their human and animal clients. However, as Kac said earlier, there have been tragic incidents in which dogs have died or been seriously injured due to negligence by a groomer, and in many states pet groomers are not licensed or regulated. You're entrusting your precious furry family member to a stranger, and in most cases you won't be there to watch over her, so you need to feel your dog is in the best hands.

The following checklist of what to look for will help –

- Is the groomer a member of a recognized organization such as The National Dog Groomers' Association of America, The Pet Industry Federation, or The British Dog Groomers' Association? Their members will have a qualification in grooming, so you can feel confident that they have studied and had practical experience before setting up salons

- Does the groomer have any dog behaviour qualifications as well as grooming qualifications? Many don't, but an understanding of dog body language and behaviour means that the groomer will pay close attention to your dog's emotional state, and adjust his or her approach accordingly. This is hugely beneficial for your dog

- Are the dogs placed on tables with hanging collars? Sadly, dogs have been known to have jumped or fallen to their death because of these. If hanging collars are used, can the groomer assure you that the dogs are supervised every moment they are attached to them? If not, or if you see an unattended dog, walk away

- Are heated cages used for drying? If so, ask for cool air to be filtered through your dog's cage instead, to reduce the risk of suffocation, dehydration or heatstroke. Insist that your dog is attended at all times so that she can be removed from the cage at any sign of discomfort

- If your dog is of a nervous disposition generally, and the groomer suggests giving light sedation, walk away. Groomers are not qualified to administer sedatives. If your dog really, seriously, needs grooming and your vet offers you a mild sedative, it's safer to ask for the medication to be administered at the veterinary surgery, and for the grooming to be done there. Dogs can react to drugs in various ways, and you should insist on staying with your dog throughout if the grooming doesn't take place at the vet's

- How are the dogs being treated at the groomer's during your visit? Do they look happy; do they seem glad to see the groomer, or do they looked scared or uncomfortable? As Kac found, Talulah was worried and looked back for Kac while being led away, and this is natural – she may have thought she was being abandoned – but she was happy and bouncy afterwards. Look at how a dog behaves towards the groomer after the session is over, and this will give you a good idea of how caring the groomer is. Is your dog affectionate and relaxed with the groomer? If so, you've found a keeper!

Kac

Whenever we have an 'errand day,' and Talulah has to go in her doggie area in the back of the SUV, we try to give her a special hamburger treat. She adores a freshly-cooked hamburger patty with a slice of cheese on top. We're non-meat eaters, but Talulah responds well to beef, so we give her these dietary treats when we are on extended outings. I'm sure Lisa will have guidance to share.

Lisa

Diet is a complex subject that warrants a book all to itself, but, in this instance, Kac is talking about a special treat for Talulah when they go out. Well, why not? Hamburger and cheese is a fatty meal, but, fortunately, it's not served up on a daily basis, Talulah thoroughly enjoys it, and suffers no ill-effects such as indigestion or bowel disturbances. She's in good health, and it makes Kac feel warm and glowy to give Talulah something that's not a part of her everyday diet.

I know dogs who'll kick up a storm for an ice cream, and are allowed one occasionally on hot days. Really, for extra-special snacks, most foods are fine, so long as they don't include ingredients that your dog reacts to, and foods that are toxic for dogs.

Here's what to avoid so that your dog doesn't become sick, bearing in mind that some dogs also have negative reactions to ingredients such as wheat and dairy.

Foods which are toxic, even lethal, for dogs

- Chocolate
- Grapes and associated foods such as currants and raisins
- Nuts
- Onions
- Mushrooms
- Hops
- Salty foods
- Yeast dough, uncooked
- Raw fish
- Fish with bones
- Cooked meat bones, especially chicken (which splinter)
- Milk (some dogs)
- Tea
- Coffee (and anything containing caffeine)
- Alcohol in any form
- Food that has spoiled/gone off

Some of the foods that are bad for dogs ...

... and likewise with liquids.

How to spot an intolerance reaction in non-toxic foods

Food intolerance is a digestive problem, in which the dog's system can't digest and process a particular ingredient. This is very different to an allergy, in which the dog's immune system mistakenly identifies an ingredient as being harmful, and subsequently produces defensive antibodies to fight the 'invader.'

The following are common signs that a food isn't agreeing with your dog. This may be an ingredient in your dog's regular food, or a reaction to a food that is given only occasionally –

- excessive scratching and itchiness, especially soon after eating, but can be prolonged
- an itchy rear end (check for worms in this case, too)
- an upset tummy, diarrhea or vomiting
- flatulence
- excessive face and muzzle-rubbing after eating
- paw-licking or rubbing
- obsessive licking or nibbling
- hives
- skin rash
- irritability
- restlessness or excessive sleeping

If you notice any of these signs, make a note of when they occur (after meals, or after an unusual snack, for instance) to try and pinpoint any obvious cause, and discuss this with your veterinarian. Like us, not all dogs thrive on the same food, and even littermates may have different dietary needs.

Kac

I have recently purchased a special tub, and a (quiet) spray nozzle to give Talulah doggie baths at home most of the time. She seems happier when bathed at home, and we have fun with the towels, drying, and shaking off the water. I use hypo-allergenic shampoo to prevent allergies, and warm, not hot, water so the experience is comfortable.

Talulah goes to the groomer every six to eight weeks to keep her coat trimmed and nails buffed. I am watching out for the next sale, as an electric home nail grooming tool just might be on the horizon so we can really do a doggy-spa at home. Or am I sniffing too much catnip?

Lisa

Home grooming is a great idea if your dog isn't too high-maintenance. Your dog will be more relaxed, and the home beauty session can be fun for both of you!

Summary

- Some dogs are low-maintenance, whilst others need regular, professional-level grooming
- You can teach your dog to enjoy being groomed
- Rewards help your dog become used to grooming

- Seek out groomers who are registered with a respected organization
- Be aware of potential safety risks
- Note which foods are toxic for dogs
- Food intolerance is different to food allergies
- Home grooming can be more relaxing for your dog ... and fun for you!

10 Talulah goes to camp

Kac

The few occasions when I thought about getting a dog usually occurred after I had been around someone who had one tucked in their tote bag: for example, when Jo Anne Worley came to a birthday party I was attending with Harmony, her sweet little Yorkie, tucked neatly inside a carry bag. Harmony goes everywhere with Jo Anne, and is a living doggie doll. This was exactly the kind of dog I wanted. A perfect angel who would sit next to me and cuddle, be easily transported, and welcome everywhere I went. I must have played with Harmony for an hour while Jo Anne and I chatted about our former cheating spouses.

A few years later, along came Talulah, and the portable tote bag idea went sailing out the window in the beat of a gnat's heart. (Those big brown eyes of hers were a magical convincer!) One of the first things I had to consider when Talulah pranced into my life, besides trashing my fantasy of transporting her around in my purse, was who might be a caretaker for her when I was away on a business trip. I spent hours researching online, looking for local kennels and boarding places. I read reviews, and paid attention to what people said, and how they said it. I narrowed my choices to a few, and made appointments to go meet the staff and survey the location (I have never been fond of the interview process).

One thing I knew for sure: I did not want to leave Talulah home alone, and have a feeder/dog walker come to visit her. She would go stark raving mad, or, as they say in the UK, barking mad, and I didn't want to leave her lonely and frightened. (We know how well that worked the day she ate her doggie mattress for lunch.)

The top-of-the-line local kennel was lovely – sparkling clean, and very chi-chi – but it was also run by focused business people who didn't show any spontaneous or natural affection toward Talulah, and seemed more interested in her health records and paperwork than her. We thanked them and went, tight-lipped, on to the next kennel. Tah tah.

Place #2 was up a steep hill. The access road was bumpy, and the car bounced from side-to-side up the well-worn single lane roadway, shaking us up like a festive martini. The day was exceptionally warm, and we were greeted by the sound of enthusiastically-barking and yelping pups of all shapes and sizes.

We were met at the office door by Jason, a smiling worker whose ensemble was a little tattered here and there, but he won our hearts when he immediately bent down to greet a reluctant Talulah, waiting patiently on bended knee until she warmed to him enough to take the treat he was offering.

We were escorted on a tour of Vineyard Kennels, so Talulah could see the 'rooms' of the dog lodging. It was explained that each dog was assigned a sleeping room, complete with elevated mattress and blankets for bedding, plus pillows and assorted cuddly stuff hither and yon. "Talulah could have her own room, or, if she made friends, she might be able to bunk in with a pal if she wanted company."

We were informed that the daily routine began at around 8am, when the lodgers were fed breakfast in their respective rooms, after which they were let out into one of four play yards, depending on their size, friendliness, and athletic prowess. They could stay out all day if they wanted to, and clamber over the outdoor furniture and man-made hills, sleep on the grass, romp and run in the dirt area, and drink out of the outdoor water tubs. Or they could go back inside.

Someone was always checking on the dogs and assessing their needs. They even showed us plans for the shallow dog pool that was in the works: we got a hearty wave from the guys already excavating the pool.

The accommodation was not The Ritz, but it was clean enough, and looked kinda fun. The inside temperature was regulated via an industrial-looking supply duct, depending on the season. All we had to do was drop her off, bring her food complete with instructions, and say adios. "She'll be fine," they assured us.

The place seemed kind of eccentric, but Talulah wasn't afraid, and even seemed curious. By the end of the tour she had jumped up into Jason's arms. We booked her first two-night stay.

I had never had to pack for a dog before, so when it came time for the trip, I found myself making a checklist for myself and a checklist for Talulah. She had a pink sweater for cold nights, so I packed that, along with her regular food and some special treats. There were instructions, a list, and a number to call in case of an emergency. I dropped her off at the kennel the next morning, and headed out to the airport for my flight.

As I said goodbye to her, there was a sad look in Talulah's big, brown eyes. Jason took up the slack and immediately began talking to her, as off she went to get settled in her room.

Two days later, I returned to collect Talulah. She was a mess. She had run her tall model legs off, she was exhausted, her tail was knotted, she was covered in a thin layer of clay dust ... and was the happiest I had ever seen her. Jason and Carson came out to say how much they had enjoyed having her, and presented me with her report card, which showed all As and the award for 'most athletic dog.'

Talulah jumped up into my arms, Miss Lickety-face. They said she had played and played, and run up and down and over and out of every item they had in the play yard. Jason told me, "She is the kind of dog that makes it all worthwhile." My mother's heart almost exploded. Not only had Talulah had a ball, but she brought joy to their world, too. The only thing we needed now was a bath.

The second time she went to stay her report card read A+ for everything, and she was noted as 'class clown.' I think we had found her perfect home-away-from-home.

Lisa

Leaving your beloved dog with strangers is one of the hardest things to do. They're family members, after all, and parents rigorously check out prospective babysitters and daycare facilities for their

Talulah, exhausted after spending a few nights at the kennel. Her report card indicated she was the Class Clown.

children to ensure their little ones will be in safe, loving hands. We need to do the same for our dogs when we plan to be away from them.

Like Kac, I would have chosen the facility that was outwardly a little messier, but where the people working there clearly cared about the dogs, and put their needs first. Our dogs don't know whether we've abandoned them forever,

so being handed over to kennel staff can be very scary for them. Introductions to a new friend there before the handover visit, and the opportunity to have a look around and become accustomed to the sights and smells, can make it much easier for your dog to settle in.

Asking friends for recommendations is a good way to create a shortlist before you begin looking around boarding kennels. This is far preferable to online reviews, which may be fake, or when a place is right for one dog but not another. The following tips detail warning signs to look for, and pointers for how to decide on an environment your dog is likely to be happy in while you're apart.

The no-nos

- You're not invited to take a tour
- Dirty or smelly kennels
- Badly-maintained kennels and exercise areas
- Evidence that clean-up after toileting isn't done frequently
- Lack of comfortable bedding. Ideally, there should be bedding both inside the sleeping area and outside in the attached run. A raised bed outside is best, as these help dogs feel more secure
- The staff are friendly towards you, but don't make an effort to make friends with your dog
- Other dogs there don't look happy or well cared for
- The noise level. There's a big difference between the barking of happy, playful dogs and anxious dogs who are calling for help
- Proof of vaccination isn't asked for – all dogs in boarding kennels must have their vaccination papers checked to minimise the risk of disease spreading
- Your dog seems scared and nervous of the staff

The 'yes' factor

- You're invited to look around, everywhere and the daily routine is explained to you
- The kennel staff pay attention to your dog, and your dog seems to like them
- The kennels are clean and comfortable, with an outside run for toileting as well as indoor sleeping area
- The bedding is clean, and the dogs all have beds rather than straw or newspaper (seriously, there are places where no real bed is provided!)
- The kennels are heated in cold weather, and have air conditioning or a cooling system in hot weather
- The dogs appear happy and relaxed, and are keen to interact with the staff
- The dogs are all supervised during outside play to ensure no bullying occurs
- There's plenty of fresh water available at all times, and the staff are happy to accommodate special diets
- You are asked to provide proof of vaccination, and any known health issues are discussed
- There is a routine for each dog, depending on individual needs. Some dogs need more exercise than others, and some elderly dogs may need smaller, more frequent, meals
- There is evidence that the staff are spending time with individual dogs in-between exercise periods
- The staff offer to send you email or text updates on how your dog is doing whilst you're away. It can help you feel happier to receive a photo of your dog having fun
- An option that is now being taken up by a lot of people, as an alternative to a stay in kennels, is the independent dog boarding facility. Your dog can stay in someone's home, or in kennels on the premises, and the people who run these as small businesses tend to put a great deal of effort into caring for the dogs who holiday with them

Your checklist for choosing kennels

can be just as useful when you're looking into taking your dog to an independent boarding facility. If you choose this option, be sure to check that the person running the business is licensed by the local authority, and is fully insured, or is registered with an organisation that takes care of these matters.

A happy holiday for your dog

Packing for your dog's holiday, as well as your own, is much easier if you make a list first – otherwise you may find yourself worrying that you've forgotten something important. Dogs don't need as much baggage as we do, but taking along a favourite bed, toys and treats can help make her temporary stay less stressful.

Some kennels will provide food for dogs, but it's much better that your dog's diet remains the same. The stress of a new environment, without you there for reassurance, can set the scene for an upset tummy, and a change of diet will make this even more likely. So, pack plenty of your dog's food, and check beforehand where it will be stored. The kennel should have plenty of space arranged for safe storage of wet, dry, and raw foods, and a labelling system that ensures your dog receives her own food and not someone else's.

Take her regular bed with you, so that she has a familiar area for sleeping. This will have her scent on it, and also residual scents from your home. It can help to also take along an unwashed sweater or t-shirt that you've recently worn. Putting this in her bed will help her feel closer to you, and she may settle more easily. Favourite toys for her kennel will increase that much-needed sense of familiarity, although it's best that soft toys are left in her kennel when she goes out to exercise. You don't want any conflict to occur over these if another dog takes a fancy to them!

If your dog is a little insecure or nervous, it can help to take Pet Remedy spray to disperse around the kennel as soon as you arrive. This is a mix of essential oils that can have a calming, soothing effect on dogs. Zylkene tablets can help, too. Zylkene is a nutraceutical which contains casein, a milk protein that has a similar soothing effect to mother's milk. If the kennel staff are willing to administer Bach Flower remedies, ask them to add two drops each of Rescue Remedy and Walnut to food and water twice daily. Rescue Remedy can help ease the shock of the new environment, and Walnut remedy can help her to adjust more easily to the temporary change in her circumstances.

If your dog wears a coat or sweater in cold weather, add that to her suitcase. If she suffers in hot weather, a cooling mat or cool coat is a great addition to the list – though, of course, kennel staff must also take steps to ensure all of the dogs are kept in comfortable temperatures.

The true litmus test for whether this is the right place for your dog is her behaviour when you collect her after the first visit. She should be delighted to see you, but also happy and relaxed around the kennel staff. If she seems cowed, anxious or subdued, do not return there. If, like Talulah, she's a bouncing bundle of joy, and the staff give you an enthusiastic litany of her adventures, you can relax in the knowledge that she'll enjoy her visits there next time you have to leave her.

Summary

- It's important to make an informed choice when deciding on a boarding kennel
- Good recommendations from friends are worth more than positive online reviews
- A checklist of what to look for (both positive and negative) can help you decide on the best place for your dog
- The happiness and wellbeing of other dogs there is a vital factor
- A clean environment where your dog is

made welcome is a good sign

- Take your dog's bed, favourite toys, food, and something you have worn with you
- Natural remedies and nutraceuticals can help relieve the stress of being left
- Your dog's emotional state when you collect her is a good way to check whether or not she has enjoyed her holiday

11 The bribe, the dog, and the dog park

Kac

I bribed Talulah. I told her that if she went grocery shopping with me, and was a good girl in the car, I'd take her to the dog park as a treat on the way home, and, as promised, after hauling the bags of groceries to the car, I took her to the local dog park for some exercise and free-range running.

Our local Dog Park.

It was Talulah's first time visiting the local dog park, and she was scared. Historically, she is not comfortable confronting other dogs while on her leash, and lunges toward them, not in a vicious way, but in an excited way, and the other dogs seem to shy away. This surprised me because, at just 11lb in weight, she is not really big enough to threaten much of anything!

Some of the owners have said things like, "What a mean dog," and pulled their dogs away. My mouth dropped when I heard this condemnation, and I followed after one such woman to explain (hopefully in a polite tone) that Talulah was a rescue, and I was trying to acclimate her to meeting other dogs.

Unfortunately, the woman had neither the time nor patience for my emotionally and physically challenged dog, and told me in no uncertain terms to keep *my* dog away from *hers*. I almost cried.

Talulah and I decided that this was not the best day for a trial run at the dog park, and went on our usual walk to get some exercise before going home.

My beginner's mind reduced this problem into a simple equation: too many new things all at once. We'd pick another time to go to the dog park when no other dogs were there, so we could let Talulah sniff and smell the place and get used to the grounds on her own.

And she had a ball when we tried out this theory. Off-leash, she ran like the wind. I remember, as a child, my

mother took me to see Rudolf Nureyev dance. He spent most of his time on stage twirling in the air and leaping like a cheetah across the African grasslands. That was Talulah: run, dash, twirl, bounce, leap. She was in heaven. She looked back to see if I was watching, which I was, in total amazement. Those long, Italian Greyhound legs transported her like lightening across the plains, with her ears plastered back against her face as she raced around the park four hundred times – or so it seemed. No wonder she was deemed 'most athletic' at the dog kennel. And she wasn't even out of breath!

Talulah can run and frolic in the wide expanse of the Cambria Dog Park ... but which way to go first?

Talulah gets on well with other dogs now. Part of that success, I believe, is due to her visits to the doggie hotel, and part of it is because I have learned to let her off-leash as soon as we arrive at the park, so that she is free to meet and greet other canines on her own terms.

There is one dog who frequents the dog park who isn't friendly, and is a little aggressive toward other dogs, and this frightens me a lot, especially as his owner seems unconcerned, or maybe even uncaring. On several occasions we have chosen to avoid him, if he's around, and just go home. There are plenty of other exercise and running opportunities, and we don't want to risk an encounter that ends up with a dog at the vet's. We still go on shopping errands followed by the dog park, but are very careful to notice who's there before we walk through the gate.

Lisa

Dog parks can be the equivalent of heaven or hell for dogs. Kac made the right decision to take Talulah when the dog park was quiet, after her initial negative experience, as this enabled Talulah to run free and have a great time, *and* get used to the scents of other dogs without feeling threatened.

Dogs are social creatures, but our domestic dogs have evolved to choose the company of their favourite humans over that of other dogs. It's a very different case with feral dogs, who have grown up among their social group and are naturally shy of humans – but even feral dogs who prefer being with their own kind don't generally welcome the appearance of strange canines. The key to whether *your* dog will be comfortable around unfamiliar dogs is early socialization. If this hasn't happened, and the critical social interaction period during the first fourteen weeks of life has been missed, a dog will find groups extremely stressful, and will need careful introductions to calm, easy-

Talulah making a new friend at the Dog Park.

going, friendly dogs in order to learn that company is a good thing.

And that's one of the downsides of dog parks: they tend to have a mix of small and large, friendly and antisocial dogs, and a negative experience (such as Talulah's with the aggressive dog at the park) can lead to fear issues that can last a lifetime if the dog doesn't receive help. Fortunately, Kac wisely avoids the park at times when this dog is likely to be there, but it must be tough on other dogs whose guardians aren't quite so on the ball and protective.

The upside of the park is that dogs get to run around freely, and can choose which other dogs to interact with. Often, groups of dogs will become firm friends, and their get-togethers will be the highlight of their day. It's beautiful to see dogs expressing their essential dogginess, with the freedom to run where they choose and to check in occasionally to make sure their guardians are aware of how much fun they're having.

Time off-leash is important, and not just because dogs can burn off some energy. Many dogs are more reactive when they're on-leash for two reasons: they're restrained, and so cannot escape unwelcome attention, and if their guardian is feeling tense or worried, this will be transferred down the leash to the dog, therefore making the dog feel more anxious. That off-leash time, though, needs to be monitored so that you can see that your dog feels comfortable and safe, and truly is having a ball.

In the UK, where I live, we don't have many dog parks, and several behaviourist friends in the USA have expressed concern that some guardians in the dog parks they visit are unaware of when their dog is being bullied, and is feeling uncomfortable or scared. It can be like bullying in a children's play area – if the mother is chatting with friends or talking on her phone, she may not notice that this is happening unless her child becomes visibly upset. So, dogs in parks need supervision, just as you would keep a close eye on a child.

It's tough on dogs to simply let loose a bunch of unfamiliar animals in a safely enclosed area and expect them to figure out relationships between them. They need to be supervised when running freely, and guardians must understand the basics of canine body language so that they can tell when a dog is bullying another ,or is being bullied. It's upsetting if your dog is targeted in an aggressive manner by another dog, and the guardian is unaware of the danger, or is aware but unconcerned. Letting them figure it out between themselves is not a good idea, as it can lead to serious injury, or worse. Understanding whether a dog is simply happily exuberant or is actually hyper-aroused, and taking action to remove an over-aroused dog before that excessive excitement leads to a fight, is our responsibility as guardians. We want our dogs to have fun, to make friends, and to feel safe.

intentions, and makes a welcome more likely.

A dog who charges full-on at other dogs and jumps all over them in his excitement may get away with such impolite behaviour during puppyhood, as adult dogs are usually more tolerant

A girl needs her rest after a busy day at the Dog Park ...

Manners maketh the dog! What's acceptable and what's rude?

Dogs have their own code of manners, and a dog who steps out of line is considered very rude, and will set other dogs' teeth on edge.

If you watch a group of dogs who have social skills, you'll notice several things. When they run towards each other, they will approach in a curving motion, or will start off running straight, and then curve away before they get too close. Curving is a clear sign to the other dog that the approaching pooch has friendly

of wet-behind-the-ears pups. But, past the puppy stage, he will most likely be growled at, snapped at, or the target dog will distance himself as far as possible. This is because a head-on charge is a precursor to an act of aggression, so dogs who lack social skills and behave in that way are considered not just rude, but also threatening.

Two dogs meeting will slow down to greet each other, and greetings are more likely to begin at the tail end than head-on. Butt-sniffing is a great way to acquire information about state of health, what the other dog has been ingesting, who she's been hanging out with (and where), and whether or not she is fertile – and that's just a little of the information those amazingly sophisticated noses collect.

A well-mannered dog who wants to play will perform an action we call the 'play bow,' whereby the front legs and shoulders go down and the butt goes up into the air. The tail wags, and the dog may give an invitation bark. If the other dog wishes to engage, she may reciprocate with a play bow ... and the game is on!

Dog Park tips

- Check out the dog park without your dog first of all. If there's a lot of rough play, and some dogs look scared, it's best to either find another park or choose a time when it's quieter
- Is the park well maintained, secure, and kept clean? It's the guardian's responsibility to pick up faeces, of course, and this is another reason why it's important to keep a close eye on your dog. If your attention is elsewhere, you may not notice when (and where) your dog has pooped
- Are the guardians aware of where their dogs are and what they're doing? A lot of conflict which can traumatize dogs can be avoided if only the guardians would step in and remove an over-excited or aggressive dog, or one who is nervous
- Is there a mix of a lot of large, powerful dogs and lots of small dogs? Big and small dogs can get on well (my Skye is a large breed mix, and one of my foster dogs was a tiny Jack Russell terrier who he was great friends with, even though she barely reached his ankles), but, again, this situation should be closely monitored so that small or nervous dogs don't feel overwhelmed
- Does your dog enjoy company? If so, a dog park can be a great place to be. If she's nervous or fearful, it's best to enlist help from a qualified, force-free behaviourist before you throw her in at the deep end with a large group of dogs.
- Is your dog well-mannered? The old school of thought was that dogs should teach manners through disciplining each other – but teaching manners is the guardian's job, and should be done *before* you begin going to a dog park. It's no fun for even an easy-going dog to be repeatedly mugged by an excited, exuberant pooch
- It's best to avoid taking toys to the park with you, as these could set up a conflict situation if another dog wants them – in the dog mind, possession (even temporary) equals ownership!

Watch the language!

Dogs communicate primarily through body language, and it's important that we humans learn to 'speak dog' so that we can decipher what our furry friends are saying. It's not a difficult language to learn, but there are a lot of subtleties, and, as you progress in your understanding, you'll start to notice these, and recognise how each area of your dog's body is involved in expressing the emotions being felt. The following is just a short list of the many signals you can expect to see before your dog makes a sound, but it will enable you to figure out whether your dog is feeling uncomfortable or frightened, as well as identify when she's having fun, so that you know if it's time to step forward and be her champion.

Uncomfortable, scared, nervous, worried or unhappy

- lowered body or cringing
- tucked tail
- ears back
- turning away the head
- closed mouth
- lip-licking and nose-licking
- rolling onto her back (a very scared

dog may urinate at the same time)
- lying down, sniffing the ground, scratching, licking the genitals (these are called displacement activities, and are intended to defuse potential conflict)
- towards humans – submissive/ appeasement grin, showing the upper teeth to create a 'smile'

Angry (aggressive)
- standing tall
- raised hackles
- a tense body – standing stiffly
- rounded, staring eyes with part of the white showing (this is called whale eye)
- a fixed gaze
- ears pinned back tightly or held tensely erect
- raised, stiff tail. The tail may wag in a jerky manner
- lower jaw and mouth tight and shifted forward (this is called the commissure; in angry humans this is evident as tightly-pursed lips)
- lip curled to display the teeth

Confident
- relaxed upright posture
- high tail
- alert expression, but not tense – no sign of whale eye
- bright eyes
- open, relaxed mouth
- ears pricked forward or loosely relaxed (not tight against the skull)

Happy or excited
- body is upright and relaxed
- Ears pricked up and forward, or floppy and relaxed
- 'Soft' gentle eyes
- open mouth; tongue may be lolling
- canine 'grin,' showing the teeth in a smile (please note that worried dogs also grin when showing appeasement or submission, but stress signals are also visible in their body language)
- tail raised and wagging
- may prance or dance in delight

Skill in accurately reading dogs comes through noticing what is happening in every area of their body, and then putting together the pieces of the puzzle. Recent scientific studies have indicated that pre-school children perceive a dog showing her teeth as smiling, so they're more likely to approach and make contact in this instance. Dogs do exhibit a form of smile, but there's a huge difference in intent between the appeasement grin of a dog who is being chastised, a tongue-lolling happy, friendly grin, and the display of weaponry that a dog uses to show her intent to bite if approached.

A wagging tail combined with soft eyes, relaxed ears and body, and a bouncing step as a dog approaches, indicates a happy, friendly dog. However, a tail that wags in a jerky motion, combined with tense posture, flattened ears and staring eyes is a clear signal to retreat before the dog launches an attack.

If your dog is being pushy towards another dog, or is showing any of the uncomfortable, anxious signals noted above, it's time to intervene and remove her from a stressful situation before it can escalate. Sadly, many dogs with fear and aggression issues towards other dogs have developed these after an experience of being bullied or intimidated, so it's essential to step in before your dog suffers psychological as well as physical damage.

Many dogs – and especially those who are well socialized – thrive on interactions with other friendly dogs, and Kac's description of Talulah's happy times with friends in the dog park demonstrates that even dogs whose social skills weren't given enough opportunity to develop fully *can* learn to thoroughly enjoy the company of the right canine companions.

The Cambria Dog Park remembers its patrons with a memorial board.

Summary

- Check out your local dog parks without your dog prior to visiting, and observe the interactions between the dogs there
- Be observant. Are dogs being bullied, or are they having a good time together?
- Think about your dog's personality, and aim to schedule park visits for times when compatible dogs are there
- Take steps to understand the basics of body language so that you can tell if your dog is feeling uncomfortable

12 Talulah and the smelly 'squirrel'

Kac

Talulah loves to be out on the back deck, which is over two hundred feet long, and wraps around the entire backside of our ranch-style house, situated smack-bang in the middle of a two-acre forest, surrounded by tall trees of the oak, redwood, and Cambria Pine varieties.

Thousands of birds live among the trees – woodpeckers, blue jays, scrub jays, crested jays, robins, hummingbirds, a rafter of turkeys – as well as butterflies, squirrels, gophers, a gorgeous red fox, a bobcat, and a myriad of other forest creatures, including the five families of deer that also think they own the deeds to the property.

It's all quite busy, and very entertaining for a curious pup. Barking and twirling and gazing and prancing, and all sorts of contortions go on when Talulah spots a new critter. Fortunately, the gated and elevated deck extends into the forest, safely supported by posts twenty feet high, and from which Talulah watches while the parade of forest critters passes by. The only thing missing is her throne.

When it's not raining (and that's pretty much every day of the year, since California is experiencing a four-year drought), Talulah spends time out on her deck communing with nature, and either talking to, or scolding, the wildlife. She's very good about letting us know when she wants back in: she uses a staccato bark as her doorbell.

When HRH Talulah is not out on the deck commanding the troops, squirrels meander up in search of yummy peanuts. In fact, we have a large gallon jug filled with lip-smacking squirrel treats that range from low-fat animal cookies, to pretzels, peanuts, crackers, popcorn, and left-over wholewheat bread: a veritable pot pourri of squirrelly goodness.

Talulah watches the squirrels through the glass door leading out to the deck, and would love to go outside to play (chase!) with them. There are certain spots on the deck where squirrels congregate for snacks: outside the dining room, outside the living room, and outside the home office. (Okay so we're a little indulgent with treats – guilty as charged.)

One summer evening, just after the sun had set, Talulah made us aware that she wanted to go out onto the deck. We had laid out a big grass pad there for her personal convenience, and thought she was headed for a wee twinkle. We heard Talulah bark in a burst of excitement, then came a sharp yelp accompanied by the sound of a major chase. We bolted

Talulah's new friend, the smelly 'squirrel.'

to our feet to check out the chaos, and watched as a frightened skunk made his way down the back stairs, after exiting the deck under the gate where Talulah could not reach him.

The scene that remained resembled bedlam. In his terror, Mister Skunk had sprayed Talulah ... and the wall, the potty grass, the plants, the barbecue, the patio furniture, and darn near everything else on the back deck! One surprised little skunk had run along the planks, spraying the whole time he was making his escape with Talulah right behind. Apparently, Talulah mistook him for a black squirrel who just happened to have a white stripe painted down his back. She had no idea that this little guy packed such a mighty punch!

It was already 9pm. The grocery store was closed, and we stood, much confused, wracking our brains about what to do. All we could remember was something about tomato juice being able to reduce the overpowering smell, but there was no tomato juice in sight; not even a forgotten can of V-8 that might have been a substitute, at a pinch.

Heading to the refrigerator, we squeezed a few salad tomatoes, and doused Talulah with the juice. Now, not only was she reeking of skunk spray, but was also dripping with ripe tomato pulp and seeds. It was not a pretty sight ... or smell.

One of us dashed to the computer to search for a remedy whilst the other ran the bath water. Talulah's eyes were stinging from the spray (and maybe also a little from the acidic tomatoes), and the bath did relieve her immediate discomfort, though did little for *l'odour de skunk*. I'd like to say that her next trip to the groomer resolved the problem, but, sadly, it did not, and Talulah smelt of skunk spray for at least three months, no matter what shampoos we used. I even bought a smell-good 'plum' spray for her, but this blend of skunk spray and plum fragrance was not a happy one.

The skunk was just helping himself to some of the left-over squirrel snacks from that afternoon's high tea. We certainly learned from this experience to ensure we switched on the back lights for a look-see before releasing Talulah outside on the deck after dark in future.

Lisa will probably have a few more thoughts to share on the subject of wildlife and domestic dogs.

Lisa

Dogs are natural predators, and even well-fed, pampered pooches retain their instinctive impulse to chase moving objects. We send balls and frisbees hurtling through the air to satisfy this natural inclination, and working dogs such as Collies are trained to harness their innate predatory instinct so that the sheep or cattle they skilfully manoeuvre come to no harm. Add a snifter of territorial rights to the mix and your dog may bark and patrol the boundaries of her property to repel all intruders. Talulah would have been very interested in the squirrels, because their fast movements made them irresistible chase objects.

Talulah's misadventure with the skunk must have been pretty awful for all concerned, including the terrified skunk, and the stench lingers. And lingers. Tomato juice is one way to ease this, but, if you live in an area where skunks come visiting, it's useful to keep a skunk odour remover handy. Other good additions to your emergency kit are three per cent hydrogen peroxide and baking soda.

If an encounter with a skunk goes badly wrong, keep your dog outside, flush her eyes with cool water if they look sore, and mix one quart of the hydrogen peroxide with a quarter cup of baking soda, and a teaspoon of liquid dish-washing soap. Wear rubber gloves, and rub her down thoroughly with the solution (avoiding her eyes) before rinsing with clean water. Follow this straight away with a shampoo and dry. Hydrogen peroxide can bleach fur, so don't be tempted to leave it on her before she has her rinse and shampoo. This mix works well, but **must only** be made up when needed because if you pre-mix it for possible future use it could explode.

If the skunk spray has transferred to your clothes, get them smelling sweet again by adding half a cup of baking soda to your washing detergent.

Dogs have a habit of rolling in things that hold no appeal for us whatsoever. Fox or cat poo perfume? Cow pat cologne? Dogs love it, though we certainly don't! When Skye was young I adopted a beautiful, elderly white Greyhound called Orla. She was what is termed a 'cruelty case,' sent to the UK from Ireland, and had suffered terribly in her ten years. She adapted to life in our home (her first home ever) as if she'd been waiting all her life for that moment, and it was beautiful to see her lounging on the sofa, playing with Skye, and eagerly running to me with bright eyes and a happy smile each time I called her name.

Orla's only less-than-appealing habit was her fondness for rolling in cat poo, and there were a lot of cats in our area who were only too happy to leave gifts for her in the garden during the night (with two Sighthounds galloping about, the cats were canny enough to stay away during the daytime). The only times Orla visited the groomer were after her far superior eyesight had spotted an inviting pile on the grass before I did, and she would hurl herself joyfully onto her back for a good roll. The smell was not pleasant! Tomato sauce can help to mask this smell, too, but, given that she had soft, white fur that had already been stained pink when she arrived, after a farmer had doused her in sheep dip chemicals, a beauty treatment was just what Orla needed.

The purpose of perfume

We wear perfume because we like the smell (and because it's considered it will make us more attractive or alluring), but dogs have an entirely different reason for getting smelly. They don't discriminate, and the more pongy something is, the more they love it. I hear a lot of complaints about dogs who don't just restrict their perfume adventures to rolling in poo: dead animals and birds, especially if the carcasses are rotting, get those poochy eyes sparkling with delight as they dive in to smother themselves in the scent.

The instinctive urge to stalk and hunt prey is still powerfully present in our domestic dogs, and it's thought that they smear themselves in 'perfume' to mask their own scent. It's also very likely that they simply enjoy smelling of something they find utterly wonderful! Another factor can be our own reaction when our dog does this, giving her a lot of attention, even if it's not the most positive sort. We may actually be reinforcing (and therefore encouraging) our dog's rolling tendency if we make a big fuss about it – best to clean up calmly and quietly!

Summary

- Dogs have an innate instinct to chase moving objects
- Keep three per cent hydrogen peroxide and a tub of baking powder in your store cupboard if skunks live in the area
- Dogs enjoy rolling in things that we find disgusting
- Dogs most likely wear smelly 'perfume' in order to mask their own scent

13 A common candy bar turns into a designer treat

Kac

I come from a big city, and so does Talulah. Ten years ago I moved from a megalopolis of 9.7 million people to a smaller community of 6000, and the true meaning of 'local' became startlingly apparent when I dropped from millions of neighbors to just a few thousand.

Becoming 'local' was a welcome experience after the hustle and bustle of Los Angeles, but it took some time to get used to. Parts of a local community are 'closed' to newcomers, and parts are wide open with enthusiastic arms to welcome newbies ... the trick is finding the open ones.

In our small town there is a cat shelter, HART (Homeless Animal Rescue Team), which we visited six times and adopted a rescue each time. There are local shops, local markets, local fairs, local sports, local parades, local picnics, and everything else you find in a big city, only much teenier. It doesn't matter if you are a retired CEO from a Fortune Five Hundred company, or a former railroad engineer; everybody pitches in at the annual events. In a small town the playing field is level.

Central California is flourishing wine country, with over 300 wineries that create some of the world's best vintages. One of the most enjoyable annual charity events is Wine 4 Paws, where most of the local wineries donate proceeds from tasting fees and sales to benefit the local Woods Humane Society. We helped with these events where and how we could, and went wine-tasting in the spirit of charity! Local people helping local animals in need of local homes.

One Friday, on the local radio talk show program, hosted by author, animal lover, and social activist Dave Congalton, the featured guests were homeless canines and handlers from the Woods Humane Society shelter. Besides bringing a few animals with them, and describing their qualities to generate a spirit of adoption in the listeners, the handlers also discussed the shelter's requirement for items such as blankets, pillows, pet food, and other staples.

Since we had a few extras of all of the above, we headed out one day to drop off some items at what we know is a very good cause. The staff were so friendly, greeting us as if we were the most amazing people in the world, and, after graciously accepting our 'stuff,' we were invited to tour the facility. "Have a look in that building," the young lady told us, indicating a structure on the left. "That's where we house our cats for adoption. "Oh no," we politely replied, "We can't go anywhere near cats for adoption because we are too weak-willed, and we already have six beautiful meowing rescues at home." "Oh, okay," she said with a smile. "Then why don't you have a look at our dog facility across the courtyard in the back?" "Okay," we agreed. We already had Talulah; we didn't need another pet; we could easily say 'no.' It seemed harmless enough.

The facility was truly amazing: the floors were spotless, the rooms clean and airy, and the happy, smiling, friendly volunteers obviously loved animals. Between the cat building and the dog building was a fenced-in yard where dogs were able to play and exercise, and also meet and mingle with potential new owners to see if they were compatible. It was a dream facility.

In the back, a large building comprised several 'wings,' separating big dogs from little ones, and quarantining those who needed medical attention:

a comfortable facility that any human wouldn't mind staying in.

The previous weekend the shelter had had a big event, Cinco de Mayo Chihuahua weekend, complete with a Chihuahua costume contest, Chihuahua races, a Chihuahua talent contest, and a host of other Chihuahua-related activities. Adoption fees were waived, and, if potential adopters met the pre-screening criteria, they could adopt a Chihuahua for only the cost of the state license fee.

As our visit to the shelter occurred on the Monday after the event, there were a LOT of Chihuahuas in the back room, many of whom were still excited after the weekend's activities. And then there was Snickers: obviously not a Chihuahua, and all by himself in a cage, looking out with sad eyes. Marlene asked him, "What's your story, little one? Why are you here?" Both of us were struck by the tenderness and longing in his eyes.

I caught the eye of the person in charge, and managed to ask a few questions about Snickers without taking my eyes off him for a moment. This little white, fluffy ball of curls with dark black eyes looked like a poster child for dog adoptions or world peace, or global healing, or just about anything that required a heartfelt response. I couldn't remember where I had seen that face before, and then it dawned in me: he was the spitting image of the dog on the can of a popular dog food!

The spiral was occurring. We were falling in love; it couldn't be helped. He was irresistible and we were tumbling down the rabbit hole.

We suggested to the lady-in-charge that we should probably take a breath, go away, think about it, talk it over, and come back after we had had some lunch.

She looked slightly alarmed at this, and advised us to put a down payment

Truffle was a little timid and skinny when we adopted him – this is his first night in his new home.

on Snickers to hold him. Marlene and I looked at each other, slightly suspicious, as she continued, "He is a really popular dog, and might be gone if you wait." Just as my cynical, big-city mind was turning to thoughts of sales pressure and pitches, a couple came up behind us, and asked the lady-in-charge about the 'white dog.'

"Oh," we said, in concert, "he's ours. We're adopting him. Sorry." I felt like we were the ones who'd been rescued. The couple were terribly nice, nodding in agreement, but followed us around, perhaps hoping we might change our minds. We didn't, and held on to Snickers like pirates with a new-found bag of bounty!

Snickers had been signed off by the vet just an hour before we met him. He had been at the shelter before, adopted, but then returned. We couldn't fathom why anyone would bring back a cutie like him. The lady-in-charge told us she thought the adopter had said she was allergic to the dog. It sounded a bit fishy to us, especially after we learned that Snickers was probably a Havanese, known for their hypoallergenic benefits. Oh well: her loss; our gain.

We took 'Snickers' out to the yard, and I asked if I could go to the car and bring in Talulah to see if the two bonded. "Of course," was the enthusiastic reply, so I dashed off to collect Talulah. Excitedly, Talulah walked with me past the cat rooms and out into the fenced yard. She loved other dogs when she wasn't on a lead, so we took her off her leash inside the gated area so that she and 'Snickers' could meet and play. There wasn't an ounce of hesitation; only cuteness and a gentle face nudge for the first meeting. After about 15 minutes of playtime, noticing the couple still silently hovering, we decided that Snickers was destined to live with us ... and Talulah agreed.

Scooping Snickers into my arms, we took him to the front desk to finish the adoption. The lady-in-charge had already

Truffle responded well to healthy food and lots of love. Here he is one year after his adoption.

notified the adoption lady that she felt we would be great parents, and were authorized to adopt the dog. With Talulah on one leash, and 'Snickers' on the other, the two of us tried not to appear in any kind of hurry to race to the finish, when, in truth, we were almost breathless.

There were pages and pages of forms to fill out; pages of agreements to read and sign; interview questions to respond to that were almost test-like in content, but very well on point and not a problem to answer. The shelter was covering all bases, and we were happy to comply. It's always fun to pass a test with flying colors, and even more so in this case as the couple were still hovering in the background, watching our every move, lest we decided not to go ahead.

"So," asked the volunteer at the front desk, "what are you going to call him?" We looked at each other blankly. Clearly, it was an important question. The name had to start with a 'T' (like Talulah) and be sweet, as he was. "How about 'Truffle?'" I suggested. Approval was unanimous, and Truffle's little ears perked up at the sound of his new name.

We'd been with Truffle for just half-an-hour, yet already loved him to bits. While the adoption lady was off making copies of things, I visited the little store on the premises and stocked up on a few supplies, including a new leash, a collar, and other accoutrements that matched Talulah's. And I bought them both some yummy dog treats for having been such good puppies.

The volunteers gathered round as we prepared to leave, posed for photos, asked us to please keep in touch, wished us well, and bid a sweet farewell to Truffle, about whom it was said had gone from "a common candy bar to a designer treat in less than an hour."

Lisa

Shelters everywhere need all the help they can get. Each dog who arrives in need of a home stacks up a hefty bill for food, vet checks, medical care, spaying or neutering, vaccinations, microchipping, and advertising, and there may be extras such as help with behavioural issues. The cost of saving a dog is far higher than the adoption fee you'll be asked to pay, and your donations of dog beds, duvets, food, toys, and financial aid really do help save lives.

Rescues run on the goodwill, commitment, and sheer dedication of their volunteers. In small, independent rescues, especially, very few (if any) of the 'staff' are paid, but their dog-walking abilities, transportation of dogs to vets or foster homes, fundraising activities, accounting skills, grooming efforts, and willing donation of time are invaluable. As many shelters are self-funding, events such as Wine 4 Paws are hugely appreciated by the shelters, as well as enjoyed by the visitors who go along for the sheer delight of a wine-tasting session. Often those dogs who enjoy social occasions are taken along to meet the public in the hope that Cupid's Arrow will strike, and potential adopters will meet and fall in love with them – which is how Talulah came to find her new home.

Sometimes love finds us

As with Snickers – soon to be renamed Truffle – it can happen that you have no plans to adopt another dog; you're perfectly happy with just the one dog in your life – but then Fate steps in and brings you the furry friend you never even knew you'd been waiting for! Talulah and Truffle accepted each other as if they were a match made in heaven, and their new companionship was made easy because they immediately bonded. Lucky pups!

It's not always idyllic, though. *You* may fall in love with a homeless dog, but your resident pooch (or your human partner) may not be too thrilled about sharing the love. It can help to have discussions with your partner before you visit a dog event or shelter. Is one of you secretly harbouring thoughts of bringing home an additional family member? If so, be honest about it early on, rather than springing an unwelcome surprise in front of shelter staff or members of the public.

The big questions

It's very hard to not fall in love when you take a tour of a shelter or meet dogs at fundraising events. All those gorgeous animals, and every one of them needs you (or someone like you) to give them the love and care they so richly deserve. This is when you need to engage your brain just as much as your heart. Questions to ask yourself are –

Truffle's eyes are full of love and kindness, now that he is happy in his forever home.

- Can you afford another dog? (pet insurance, vet fees, food, toys, bedding, etc)
- Does your resident dog enjoy having canine visitors to stay? This isn't the same as meeting and greeting other dogs while out. Ask yourself whether your dog likes having other dogs visit your home, which is, after all, her territory. If so, that's great. If not, please consider the impact on your dog of having an unwanted permanent 'guest' around. Imagine how you would feel if your partner said "It's such a joy having them around, and we have enough space, so I'm going to invite a friend to come and live with us." You may not be too thrilled! Some dogs thrive on company; others prefer to be the only dog. And even with sociable, happy-to-share dogs, the match has to be right.

  Sometimes, two seemingly mismatched dogs will become close friends, but there are plenty of households where one dog is perpetually nervous of the other, where bullying takes place, or where warring dogs have to live in separate areas to ensure they never meet face-to-face. This isn't conducive to a relaxed life for anyone concerned, and it's not a situation the rescue shelter would be keen on establishing.
- Are both partners (or all family members) committed to caring for two dogs? Very often, the responsibility tends to rest with one family member after a while: are you prepared for this possibility?
- Do either or both of your dogs need a great deal of training or behaviour help? Dogs learn from each other, and can learn undesirable habits as well as good ones. If your dog has a certain issue, such as barking hysterically when the doorbell rings, are you prepared for this to possibly be taken up by the new dog – or vice versa? And are you prepared to call in (and pay) a qualified professional to help you teach both dogs the behaviours you want from them?

Moving along

If your answers all strongly indicate that it's a good time to bring home a new family member, it can help to discuss your particular likes, needs and lifestyle with the shelter staff who deal with rehoming the dogs. This helps to narrow down available dogs to those who fit your 'picture,' which may include size, breed or breed mix, age group, exercise needs, gender, and any other factors, such as whether the dog is already housetrained.

Introductions

So, you've met some dogs and one seems perfect. You're in love, and you hope

your resident pooch will be smitten, too. The initial introduction is very important, because it can set the tone for the future relationship.

The best way to bring together unfamiliar dogs is in a neutral space, because that reduces any possibility of the resident dog feeling uncomfortable about a stranger suddenly appearing on her territory. Rescue shelters often have a space for meetings, such as an enclosed paddock or yard, and this worked very well for Talulah and Truffle. If you're introducing a new dog into your home without a prior meeting, it can ease any potential pressure if you start off by taking both dogs for a walk before bringing them indoors.

Fortunately, Talulah and Truffle liked each other on sight, and it's wonderful when this happens. If you're unsure about whether or not one dog will accept the other, it can help to have both dogs on-leash, with a handler for each dog. Take a stroll, with the handlers on the inside and the dogs on the outside of the group, so they each feel protected, and if they appear keen to meet (watch for curious glances, bright eyes, wagging tails, and a spring in the step), you can switch positions so that the dogs are on the inside, walking beside each other, with a few feet of distance between them. If they continue to look comfortable, let them greet with the usual butt sniffs for three seconds at a time, with breaks in-between, to avoid putting pressure on them.

When you feel confident that they want to interact in a friendly manner, release them from their leads to hang out together. Play bows are a great sign of future friendship, so look out for these!

If one dog appears anxious (low tail and body, pinned back ears, staring eyes or sideways looks), move apart a little and carry on walking until you see signs of relaxation. It's better to take it slowly than try and persuade dogs to make friends, because pressure only creates tension. It took an hour of walking around the nearby field for one of my very nervous foster dogs to relax, and want to engage with my dogs, Skye and Charlie. Most dogs will make a decision long before then but give them as long as they need.

A new family member

Once you're sure that you've made the right decision (I know, it's likely you knew in the first moment, but now you have considered your dog's feelings, too), it's time to sign the adoption agreement, decide on a new name if you want to change this, and take home your new family member.

Most rescues have stringent clauses in their adoption forms. If a dog is too young, or hasn't yet been spayed or neutered, you may have to agree to provide proof that you have arranged this at a specified time in the future. You'll also need to agree to return the dog to the shelter if, for any reason, he can't stay with you. This protects adopted dogs from being passed on to inappropriate homes, or even given away to be used as bait dogs. Sadly, this happens. Some rescues have short term veterinary insurance in place, so you'll need to arrange your own insurance as soon as possible after taking home your new dog to ensure that cover does not lapse.

It's always a happy moment for the shelter staff when a previously unwanted dog is waved off to start his new life . And remember: by adopting you are actually saving TWO dogs – your new family addition, and another dog who the shelter can now take in to fill the space he left.

Isn't that great?

Summary

- Shelters appreciate any help you can give
- Sometimes the right dog finds you – even when you're not looking!

Truffle settled in and nesting in a pile of yarn that matches his winter sweater.

- Ask yourself some questions when considering adding to your family
- Relaxed introductions in a neutral environment help both dogs get off to a good start
- Observe both dogs' body language during introductions
- When you adopt one dog, you're actually saving two!

14 Accidental BFFs

Kac

When we brought home the dogs for the first time, we wondered how it would all work out. Food, treats and water were no problem at all. Talulah was crate-trained for night-time sleeping, but we had no idea about Truffle. Would he howl at night? Would he miss someone or something? Would he be frightened? Was he house-trained? What should we expect during his first night in a new environment?

It occurred to me to keep both of them on the leash as we walked around the house and introduced Truffle to his new home. He was not at all fazed, and mostly walked calmly around the house, room by room, and sniffed. Talulah, who didn't let him out of her sight, walked right beside him, and kept looking over her shoulder as if to check that he was alright.

We showed him two places to potty: out on the deck (which had a section of artificial turf), and outside the front door in some garden beds and walkways. We had no fence, so potty breaks were always on-leash for Talulah, and now for Truffle, too.

I added a few drops of Rescue Remedy to the water bowl to help calm his nerves, though it turned out that this was more for my benefit than his as he showed no sign of anxiety. After about thirty minutes we let them off their leashes to explore the house together, with one of us sneaking a peak around the corner. Truffle was also being introduced to six indoor cats in kitty-land, none of whom had ever forgotten their Egyptian royal ancestry. Truffle was so calm and sweet that he braved sniffing the cats: he didn't care about 'cattitude,' he was just being friendly.

The first night at the house was remarkably peaceful. Truffle crawled up onto the sofa to watch some television; ate his dinner like a champ, and, when it was time for a potty stop, walked on his leash and did his business right beside the ever-vigilant Talulah.

Talulah became Truffle's big sister, guardian angel, and constant companion. If there had been a doggie Facebook, they would have 'friended' each other immediately!

The extra traveling cage we had for Talulah easily became Truffle's night-time crate (it was pink, but he didn't seem to mind). He had a new soft and scrunchy bed, which he snarfled down into with ease.

Talulah always loved the night-time ritual of seeding her crate bed with

Truffle and Talulah sit together, walk together, lean on each other, and nap side-by-side. They are best friends, and are inseparable.

beef roll pieces. The routine was: a quick twinkle outside, followed by entering her crate and the enthusiastic discovery of treats. Then it was lights-out for puppies, and we didn't hear another sound until morning. Truffle settled right into the routine, following Talulah's every move. He trundled along behind her every step of the way, and they were nose-to-nose in their individual crates before 10pm.

We never expected such an easy transition, becoming a two-dog household without pain, muss or fuss. It was as if Talulah and Truffle had been littermates and grown up together from day one. They were very amusing when they went for walks, shoulder-to-shoulder. Talulah has long model legs, whilst Truffle has short, stubby pins, which means he has to walk twice as fast to keep up with her, but he does so with glee. They separate to make a twinkle stop, but otherwise are linked together by an invisible thread. They fit together as tight as a marching band in a parade.

And so it continues. Talulah and Truffle leap into their beds in the back of the car when we go for a trip, their harnesses connected to interior car leashes for safety. They hunker down in their beds and usually go straight to sleep. Talulah no longer whines: her buddy is right there next to her, and she feels safe.

Truffle watches out for Talulah, too. They wait for each other, they share food, they switch beds when they feel like it, and only once did they ever growl at each other ... when a juicy beef bone was involved, and one forgot whose bone it was and trespassed. Oops!

They do everything together. They take baths together, they go to the groomer's together, and they are never more than six feet away from each other at any given time.

The bond they enjoy seems predestined; it must have been in the stars. It is simply one of the most magical experiences I've ever had; to this day, I am grateful, amazed and delighted at how very close they are, sharing what appears to be a deep and genuine love and lasting friendship … until, of course, a beefy bone becomes involved. Well, nobody's perfect!

Lisa

It made my heart sing when I heard how smoothly Truffle settled in, and how close the bond was between he and Talulah, right from the start. I've witnessed this with Skye and most of my fostered and adopted dogs, and the presence of a relaxed, welcoming dog can make such a difference to the wellbeing of the new family member. Moving home is a scary business for dogs. Rescue dogs often find kennel life very upsetting after living in a home. The constant noise, the strange environment, new routines, reduced one-to-one human company – these are all hard things to adjust to. Plus, dogs are social creatures, and (other than outdoor-living working dogs) have been deliberately bred over thousands of years to generally prefer the company of people to that of other dogs. They can feel lonely and confused, and kennel stress is common. This can show itself as the animal shutting down and becoming quietly depressed, or through repetitive behaviours such as jumping or circling, or constant barking, which is a call for help. Working in rescue can take an emotional toll, and it's always a cause for celebration when a dog departs for his new forever home.

Some dogs find it hard to accept the presence of an 'intruder' in their home, and may take a while to realize that two can play better than one, and make friends. Other dogs simply love company and welcome their new companion with open paws. Kac did exactly the right thing by taking the time to allow Truffle to explore on-leash, with Talulah acting as his guide.

Giving your new dog the

We celebrate everybody's birthday. This is Talulah on her 5th birthday.

opportunity to have a good sniff around and become familiar with the environment is vital, as is heading for the toilet area(s) so that he quickly learns where the acceptable elimination spots are. When a new dog arrives at my home, I let both dogs off the leash into the garden, after the outside introductions, for an exploring session. They naturally toilet while they're out there, so get lots of praise and a tasty treat. When they come indoors they have a good mooch around together, sniffing everything in sight, and, before long, either one or both of them will choose a resting place and relax a little, while keeping an eye out to see what happens next.

Treasured resources

Food, chews, and toys are very precious to dogs. Food is necessary for survival, of course, and chews, bones, and toys are considered to be 100 per cent the property of the current keeper (whether or not these belong, in your eyes, to a specific dog). In a dog's eyes, possession is nine-tenths of the law, so if something is left lying around, he will naturally claim ownership. This can cause conflict if the object concerned is considered equally valuable by both dogs, so prevent arguments by keeping a close eye on who has what, and removing any goodies that have been left lying around. Most dogs are happy to share, but occasionally (as with Talulah and Truffle and the bone) some things are just too special to be considered common property. Some dogs who have experienced deprivation in the past may need careful management around resources – especially food.

It can help to keep toys and chews out of the way, especially during the first two weeks, and to feed both dogs in separate rooms, with a safety gate in-between and the door open, at first. This means that if one dog has a tendency to guard food and displays aggression if the other approaches his precious bowl, you can keep both safe. Of course, if they've clearly bonded from the start you may not need to do this – but it's best to check out responses around food and toys first, to avoid the possibility of conflict that could damage a good relationship. Once it's clear that both dogs are comfortable about eating together, consider feeding them in different areas of the same room. It's always more comfortable to have some 'elbow space.'

In some cases the presence of a companion can encourage a reluctant eater to tuck in. One of my foster dogs, a dear little elderly Jack Russell Terrier called Tilly, was brought back to me when she became terminally ill with kidney failure,

and had refused all food for a week. Tilly worshipped the ground Skye walked on, even though she only came up to his ankles, and the hope was that being with Skye might prompt her to eat something.

When she arrived at our house she was so happy to see Skye that she wagged her tail for the first time in a couple of weeks, and Skye was delighted to have his friend around. I prepared a meal for Skye, and a very small, tasty meal for Tilly (large quantities of food are very off-putting to a dog with no appetite), placed the bowls side-by-side in the living room, and went to sit over the other side of the room so that Tilly wouldn't feel under pressure. To our delight she firstly watched Skye begin his meal ... and then ate hers!

And, of course, they must have matching Christmas sweaters!

Resting places

It's best to have separate beds prepared for both dogs in quiet areas, and to encourage your new family member to test out his bed for comfort once he's had a good look around his new home. Some dogs prefer to have their own space, and may end up happily exchanging beds at random, while other dogs like nothing more than to sleep tucked up against their companion. One thing that can cause amusement for guardians is how often, when there's a large and small dog, the small dog will end up in the biggest bed some of the time, whilst the large one squeezes into a smaller space.

The bed should be a place of sanctuary as well as rest, and it's best to make a rule to let sleeping dogs lie undisturbed. I've been called out to help a lot of families whose dogs are perfectly well behaved most of the time, but who will snap or even bite if someone strokes or kisses them while they're asleep. Imagine how you might feel if rudely awoken by someone manhandling you – you would probably jump several feet in the air and swat the intruder, too! The Do Not Disturb rule should be spelled out very clearly to resident or visiting children, because they find the cuteness of sleeping dogs hard to resist, and the fallout from a loving hug can be disastrous for both the dog and child if the dog reacts aggressively.

Emotions

Recent research has confirmed that dogs experience a wide range of emotions comparable to our own. In the old days, it was considered anthropomorphic and unscientific to apply human emotions or characteristics to dogs, even though naturalist Charles Darwin clearly catalogued descriptions of emotions he had observed in dogs and other animals. In his ground-breaking book *On the Origin of Species*, Darwin describes similarities in facial expressions that register emotion in humans and animals, and further tells how he recorded seeing a variety of emotions in human and non-human animals in his book *The Expressions of the Emotions in Man and Animals*, which was published thirteen years later, in 1872.

Happiness, sadness, grief, fear, disgust, anger, surprise, jealousy, and love are experienced by our dogs as well as by us, and the emotional intensity of these vary according to the experience, and response to that experience. Some dogs are naturally happy-go-lucky, while others seem to have a less positive (shall we say), more Ee-yore-like approach to life. In utero experiences, the early weeks of life, and subsequent character-forming experiences, have roles to play in outlook and attitude.

The relationship between your two (or more) dogs needs time to develop, and you can encourage this by rewarding all happy, harmonious, and calm interactions. Try to share your attention equally, when possible, without setting up a situation in which a shy dog feels under pressure.

Although some dogs are quieter, and others more demanding, situations can arise in which one experiences jealousy, and becomes pushy, snappy or withdrawn when the other dog is receiving attention. Charlie, my feral dog, needed a great deal of help to cope with domestic life and the presence of humans. However, once he discovered how good it felt to be loved and receive strokes, he became a very cuddlesome dog who loved to be in constant physical contact with me – lying with his head in my lap; sitting leaning against me, or jumping up to offer his paws for a dance lesson. For a while, Charlie resented Skye receiving any attention, and would butt in and use his body to block access to me if Skye approached. I countered this by turning away and putting myself in-between Charlie and Skye, so he would learn that, sometimes, he needed to step back a little. He quickly figured out that there was enough love for both of them, and decided that he was happy to share.

Summary

- Moving to a new environment can be scary, even when your dog has been in rescue kennels
- Allow your new dog time to explore
- Take your dogs to the toilet area as soon as possible after arriving home
- You can reduce the likelihood of resource-guarding at the outset by ensuring no toys or chews are left lying around
- Prepare separate beds, but don't be surprised if your dogs swap and change
- The bed should be a peaceful sanctuary
- Dogs experience a wide range of emotions

15 Doggie routines … and hurt feelings

Kac

As Talulah and Truffle gradually settled in with each other, I noticed that they established distinct routines. Our usual morning ritual began with getting them out of their crates once the cats were fed, and calm had settled over the household once again. From their crates on the landing, the two ambled through the living room, and out into the sunroom where we let them frolic on the back deck (as well as have a morning wee). On cue, they did their business, and usually then wanted to immediately come back into the house for breakfast … usually ours.

While Marlene and I had a morning chat and ate our breakfasts, Talulah and Truffle watched us like hawks. Only occasionally did they succeed in getting a piece of egg; mostly they just gazed longingly at us with a 'please-help-me-I've-never-eaten-a-morsel-in-my-life' look on their agonized faces, while we attempted to enjoy a guilt-free breakfast. Only after an award-winning performance of pathos did they pad on over to their full-of-fresh-food dog dishes.

Not a day went by when we didn't experience the Sarah Bernhardtian drama, which became our fond little ritual. Both dogs perfected their routine each day, and now we have a virtual three-act play every morning. It's actually become quite fun.

When night-time rolls around there is less drama, but more ritual, and this begins by cutting up little sausage treats, and placing these in their cozy doggie beds inside their crates. (I always shut the crate doors after depositing the treats, because I have discovered that the cats – who like to jump the baby gate for a foray into puppy-land – also like these, and have been known to help themselves if the crate doors are left open while the rest of the evening routine plays itself out.)

Once the dogs hear us open the treat jar lid, they are on the alert for the next stage of the plot. After strategically placing the treats in the crates, we let the dogs out onto the back deck for the final evening wee. (Motivated by the treats, it's always a fast last act.) Enthusiasm runs high as I lead them towards their crates, by way of the den, through the doggie gate and up onto the landing. Their little feet tap dance as they wait for me to open the latch and let them inside their crates. Happy campers gobble their minced sausage roll, then settle down for beddy-bye, with identical doggie smiles and happy hearts.

One night, I accidentally upset the apple cart. As per the usual drill, I cut up the sausage roll and put it in the crates, but, just as I was closing a crate, Truffle bolted through my carelessly-latched doggie gate and went straight into the open crate for the treats.

"Noooooooooo!" I was not angry with *him*; mostly, I was angry with myself for leaving the gate loose, and not him, but, at the sound of my agonised wail, Truffle immediately stopped eating his treats, and cowered in the front of the crate, looking up at me with enormous, wounded-heart eyes. Seeing his hurt little face, I was mad at myself for over-reacting. Poor little baby boy. His heart was broken, it seemed.

Scooping him up in my arms, I reassured him with "There, there, it's all right now. You're a good boy," then put him back in his crate. But he remained at the front of the crate, unwilling to move into his bed, continuing to look sad, and refusing to go anywhere near his treats.

In that instant, I felt like I had ruined my darling lad.

Then something occurred on me: maybe it was not so much my reaction as the fact that we had disturbed the usual routine. Lifting Truffle from his crate, I returned with him to the back door, and encouraged him to do his evening business. Although more hesitant than usual, he finally performed.

Truffle didn't then *run* to his crate, but, rather, walked to it at a fast pace; only then was he comfortable enough to eat his treats and lay down in his soft bed.

I felt so badly about my gaffe, but was encouraged by his resilience, and the comfortable return to his routine. I experienced a great lesson in canine behaviour, and how my overly reactive response had affected my lovely little boy.

Lisa

Dogs, like people, thrive on routine and ritual, as it helps them to feel more secure, because routines create an element of predictability as well as break up the day. Too rigid a routine can become a problem, because your dog may come to expect that certain things always happen at specific times, and he could become uncomfortable or antsy if you're a little late with something – it helps to have some flexibility.

We naturally set up routines for our dogs. We take out our puppies to

Truffle's feelings were hurt, and he took to his bed for comfort ...

... but after Kac apologized to him they were, once again, one big happy family.

toilet after they've eaten, slept, or had a playtime to teach them house-training. As creatures of habit ourselves (and with busy schedules), we tend to set walk times for before and after work. We have our meals at fairly regular intervals, and serve up our dogs' dinners at certain times, too. We may have scheduled times for games, and for training classes. Our dogs pay close attention to us and what we do, and synchronize their body clocks in accordance with ours.

Dogs have excellent body clocks, and woe betide you if you always, always serve dinner at 5pm, but then are held up for some reason. You'll most likely to be treated to the sad gaze, the pacing, the pointed looks at the food dish or refrigerator, possibly a little light whining, until you get the message and prepare the meal. For this reason, too rigid a routine isn't helpful because changes to it can make your dog feel anxious.

The same goes for walk times. If you always take your dog out at 7am you can bet your breakfast that he'll be waiting by his leash or the door just as the minute hand turns over. Life gets in the way some days; the phone rings or friends arrive, or you simply forget the time. It's best to have a flexible routine, with mealtimes or walks sometimes a little earlier or later than usual.

Like us, dogs have rituals. We may shower and brush our teeth before breakfast; they may turn around three times while they look for the best spot to lie down. Rituals help us feel safe and comfortable, and only become a problem if they become compulsive and cause us to become stressed if we can't follow them.

Talulah and Truffle watch Kac and Marlene like hawks while they have their breakfast, ignoring their own food until afterwards, because they've learned that, just occasionally, some of that tasty food will come their way. It's hope, based on occasional rewards, that keeps them close by at mealtimes. Fortunately, Kac doesn't mind this intense canine scrutiny: it gives her pleasure, and has become part of her daily routine, so isn't a problem.

I visit a lot of families, however, who aren't happy about their meals being disturbed by sad-eyed drooling dogs who stick like velcro to the dining table until dinner is over – and may even jump up to help themselves. If this is a habit you want to break, the easy way to do this is to refuse to pay any attention to your dog when you're eating, and never, ever again feed him from your plate. Instead, place a chew or something tasty on your dog's bed and invite him there. This habit can be broken surprisingly quickly if you can be strong enough to allow no-one to ever feed your dog during human mealtimes. If the behaviour (begging) is never rewarded (with food from the table), it will fade away. Relent and toss him some tasty morsel even once, however, and your dog's hope (and begging behaviour) will quickly be reinstated.

Kac's night-time ritual with the sausage treats is a lovely way to help Talulah and Truffle enjoy their bedtime routine, and quickly settle down to sleep. I give Skye a small treat on his bed, and a goodnight kiss every night, and many of my friends do the same with their dogs. It's as nice for us as for our dogs.

Impulse control is hard for dogs. They have to be taught to wait for good things (we have to teach this to our children, too), so Truffle's sudden dash through the open gate to snaffle up the treats a little early was a perfectly natural thing for him to do. In his mind, he had the opportunity, so why not take advantage of it? Kac's stern reaction would have confused and frightened him, and he wouldn't have understood why he was being told off for eating his treats earlier than usual. His shocked expression and cowering were partly because he didn't know why he was being chastised, and also partly appeasement behaviours aimed at letting Kac know that he meant no harm.

It's a horrid feeling when a beloved dog looks and acts as if his world is falling apart (which it is when we're angry with him). Kac did exactly the right thing by taking Truffle outside and then following the usual bedtime routine. This was a great way to defuse the hurt and upset, and to help him feel safe and secure enough to settle back into the usual routine, enjoy his treats, and fall asleep. It also meant that there wouldn't be fallout in the future. Often, if a dog has been told off for something, in Truffle's case for going into his crate, he may be reluctant to approach the crate in future. Repeating the familiar routine and pairing this with positive feelings helped Truffle shrug off any stress he may have been holding on to.

Sometimes, we react without thinking. A shout or sudden movement can startle a dog, so one of the many things we dog guardians can learn through this special relationship is to monitor our own emotional states and responses, and consider how these affect those around us – human and canine. The dog who taught me the most in this respect was Charlie, the extremely fearful, unsocialized Romanian feral dog who came to live with me in 2013, and I learned a great deal about myself *through* him. Because the slightest movement or sudden/loud noise would send Charlie into a state of panic, I had to keep my movements very slow and my voice soft at all times, and I asked visitors to do the same. This need to be constantly aware of how I was expressing myself through my body movements and tone of voice prompted the realization that most of what we do is automatic and unconscious. Being mindful can transform our relationships, and Charlie blossomed into a dog who brought tremendous joy, not just to me but many others, too.

Summary

- Dogs, like us, find routines and rituals soothing and reassuring
- Truffle's dash for the treats was a natural behaviour, as dogs are opportunists
- Kac's strong reaction prompted Truffle to display appeasement behaviours
- Returning to a known routine can de-stress an anxious dog
- Living with a dog can teach us to monitor our own behaviours, and consider how these impact on those around us

16 From separation to integration overnight

Kac

In the interests of sanity and safety, I felt I had to separate the cats from the dogs as soon as I brought the latter into the house, so established boundaries between the two species. I wanted to respect the cats, their age, and the noteworthy achievement of having the six of them living together in relative harmony, whilst at the same time protect both them *and* the dogs from each other!

The baby gate which separated felines from canines was a good start. Only one of the smaller-boned cats could slip between the rails, and she was not the type to start fights, so we were good.

Occasionally, we would open the gate and let the furry people mingle and get to know each other. Percival, the oldest male cat who lives to eat, took up the practice of joining us for breakfast, getting in line with the puppies for treats. Socrates, the youngest male cat, just wanted to be friends with everyone, all the time, and didn't mind if he rubbed his chin on a person, another cat, or a dog. He is such a happy little soul. Jazzmine, the oldest female, is a stevedore, and rules the lair with the authority of a disgruntled teamster captain. She is not one for socializing or making friends (she felt she already had plenty!), until there is a can of tuna involved and that gets her attention. The rest of the cats – Mandy, Harmony, and Merlynn – are observers and nappers, who couldn't give a rat's whisker about anything beyond a clean litter box and a fresh meal.

And so life continued with occasional visits between 'lands,' and a moderate sampling of what dry food was like in the opposing camp.

Then the unimaginable happened – we decided to sell the house and move 250 miles to the south.

The house we were living in had one story, and was laid out in wings like a ranch house. The new house was a two-level design, with the bedrooms upstairs and living space and kitchen downstairs. Where we would put the baby gate became the breakfast topic of conversation for weeks, but, since the new house featured bannisters with six inch gaps, no cat worth his or her salt would have any trouble getting through these to penetrate enemy lines. What to do?

My brilliant partner-in-crime blurted out one day, "Nothing. We won't do a thing."

I replied with a startled expression, "But how will we separate them?"

The cats and dogs became integrated after Truffle shared a can of cat food with the cats ...

"We won't. They'll just be integrated," was her reply.

And that was the end of the conversation. One of us trusted them to work it out – after all, they'd known each other for over a year, and had had close encounters of the food-related kind – so it should be a natural thing. But I was deeply concerned about the 'what ifs,' and fretted late at night when no one was looking.

The decision was made to move

... and now Socrates and Truffle very often snooze together.

the dogs to the new residence one day earlier than the cats, to give them the chance to familiarize themselves with the backyard and grounds, and sleep for a night undisturbed in their cozy, familiar crates. The cats would arrive the next afternoon, and be placed in one room with food, water, and litter, to adjust to a controlled environment before being let out into the rest of the house where they and the dogs would be instantly integrated.

A little butter on the cats' paws would keep them busy and content; some Rescue Remedy in their water bowl, and we felt we were going to nail the move. This was to be a brave new world for sure, but at least one of us had confidence it would work.

Only, the best-laid plans rarely leave room for the mistakes of others, and what actually happened was that all hell broke loose.

The contractors we had hired to improve the upstairs bathrooms were way beyond their contract time, and everything was dusty, dirty, in a shambles, with construction mess everywhere. We had a contracted, iron-clad 'finish date' which they cavalierly ignored, and left

us in chaos and filth. The dogs had to be protected from the dust, and couldn't roam free because of the sharp tools and supplies lying around. Nothing could be moved into any of the rooms because of work items carelessly left about every which where, and the fine dust from chiseling, plastering, sanding, and dry walling permeated every nook and cranny of the house.

We couldn't stop the train. It was on track and headed south at full speed. When we realized that the construction mess was not going to be cleaned up in a day, we couldn't derail the plans. Movers were arriving in hours to load the truck, and we were bound to keep to the dates. The cats were rounded up, put into containers and driven south by me. Four hours-worth of cats telling me how this trip was unnecessary, beneath them, and annoying was the least of my worries, although the biggest of theirs, apparently. Three of them howled their discontent for the looooong trip while the other three plotted their revenge.

When I arrived with the six felines in tow, it was not the pretty sight we had envisaged; nor was it the smooth, clean easy move we had meticulously planned. It had become the nightmare of all nightmares: the deep, visceral fear of everyone who has ever moved or hired a licensed but careless, thoughtless and unhelpful contractor.

Two of our cats instantly became ill from inhaling the dust and dodging the debris. The dogs were on edge, and none of us slept for a week. I will spare you some of the more grisly details, but it took the contractor over a month of hit-and-miss work to finish the job, and two months after that there was still dust and grime in every corner and crevice of the house.

However, when we did get enough of the dangerous sharp edges of construction material out of the way, we opened the doors, and poof!, the animals were integrated. The cats gingerly padded around the house exploring one room at a time, and were so busy discovering the new smells and dodging piles of dirt, that they didn't even notice the dogs, and acted casual and nonchalant when encountering a canine in the hallway.

This was NOT the plan we had in mind, but Fate stepped in and said, "Oh, heck no. Let's do it MY way." To this day there is tranquility, and bonds are beginning to form. I don't think they'll be chumming up to go to the movies together anytime soon, but at least the couch is shared, and certain felines like the dog food better than their own grub. We do make an attempt to keep them out of each other's dishes, but those tricky cats are slippery, and lurk in the shadows, awaiting an opportunity to sample a bit of the dog-food-du jour.

And the dogs? Well, they've been known to lie in wait for an empty cat food can, and a soupçon of mackerel residue on a cat plate. The cats and dogs are doing fine. They appear happy, and, aside from a little unarmed policing, all is well in the new abode. Integration works!

Lisa

Introducing a dog to a home that's already owned by a cat *can* be very straightforward ... providing you go about it the right way. If your dog has come from a rescue, she may have been cat-tested (introduced to a cat in a controlled manner) before adoption, but not all rescues have the facilities (or available dog-savvy cats) to do this. Plus, a dog who has been tested and pronounced cat-safe may not respond as well with a different cat, and dogs who happily accept the cats they live with may nevertheless consider neighbouring cats to be fair game.

Dog and cat introductions

When you bring your dog home and prepare for the initial meeting with

Talulah and Percival: happy to hang out together.

his feline companion, you'll need to have two adults and take some basic precautions. Remember to consider your cat's wellbeing. Is she scared of dogs, or does she ignore them? A cat who tries to bolt for freedom may spark the dog's chase instinct, so make sure your cat has been fed so that she's more likely to be relaxed, and ensure that she is held safely and comfortably.

Exercise your dog before the introduction, but don't include any exciting runs or games, because you don't want your dog to arrive wired and hyper. Keep your dog on a long leash and hold this as loosely as possible (any tension you're feeling will zip down through the leash straight into your dog). Have large quantities of high-value food in your pocket or a pouch.

Ask your dog to sit, and offer rewards for calm behaviour so that he's focused on you before your helper brings the cat into the room. When the cat is brought in, let your dog glance at her but keep the main focus on you, and offer lots of food rewards for looking at you. The aim is to keep it casual between the two animals.

If the cat is comfortable sitting on your helper's lap, that's great, but if she wants to move around then let her, whilst ensuring that your dog can't leap forward and jump on her. Avoid her becoming anxious and running around trying to find a way out: ensure there are escape routes available for her in case she needs these.

Watch the body language of both animals, and once both seem relaxed and comfortable, let them approach each other for an introductory sniff. Call your dog to you after a moment, rewarding him for complying. Be prepared to move them apart immediately you see any signs of tension in either animal.

Keep the initial introduction brief, and end on a positive note before separating them. Thereafter, careful daily introductions can promote a gradual easy acceptance of each other. Take precautions even after they're friends, though; avoid leaving them alone together, and put them in separate rooms when you leave the house.

Low-stress moving

Moving house is considered (unsurprisingly) one of the most stressful events in our lives, and moving house with furry family members in tow can make for even more sleepless nights. Kac and her spouse coped admirably, despite the chaos of the unfinished building work, and it helped a lot that their cats and dogs had already had numerous uneventful introductions over a period of time – with the delightful distraction and reinforcement of food paving the way to mutual acceptance.

There are so many potential worries around moving with pets. Will they cope with the change of environment? Will they miss the previous home? When you have both dogs and cats, there's also the issue of whether the cats will like their new home enough to stay, or whether they'll head for the hills at the first opportunity.

Hopefully, your move won't be the nightmare experience that Kac described, and your home will be ready to make your own when you arrive. It can reduce stress levels if you put one person in charge of the dogs and cats (and, if you have enough helpers, one person for the dogs and another for the cats is even better) so that you know they're being looked after. Keeping the animals in separate rooms, with the door closed and clear instructions to removal workers that animals are contained there, can prevent worries about any furry friends escaping.

Once you've all been transported to your new home, ensure that the area is safe and gates are closed, and then release your pets so that they can explore. It can help to have their beds set out as soon as you arrive, and objects that smell familiar placed around the house. You can even rub your pets with a towel and then rub this on various points around the house to spread their scent, and help them feel more comfortable in their new surroundings.

Cats will need to be kept indoors, ideally for two weeks, after a house move to reduce the possibility that they may try to return to the previous home. Wait until they seem settled and happy before you let them go outside. If your cat is used to wearing a harness and leash, you could take her out on that so that she can become accustomed to the new sights, smells, and sounds.

Dogs enjoy exploring, so, as long as your boundaries are secure, a trip into the garden as soon as the removal people have left is a great idea. It gives them chance to take in their new territory, and shows them where the toileting will take place from now on. If the boundaries aren't secure, take your dog outside on-leash until this has been remedied.

And relax!

The more relaxed you are around your pets, the easier life will be for them. Once you're in your new home and the essentials are unpacked, pour yourself a glass of something nice or put the kettle on, sit back among the boxes, and focus on helping your furry family members become familiarized with their new home. Give them space to roam around and take in the new smells, and hide some tasty treats in places where they can find them. A little feel-good time will be what you all need to help you unwind!

Summary

- Cat and dog introductions can be straightforward if you go about this correctly
- If you are adopting a dog, ask the rescue centre whether she has been cat-tested
- A dog may be friendly with resident cats, but less so with neighbouring cats
- By following the safety guidelines above you can help your dog and cat accept each other
- Ask someone to take charge of your pets during a house move
- Rubbing your pet's scent around the new home can help her feel more settled
- Cats should be kept indoors for two weeks after moving house, unless you take them outside on a harness and leash
- Check boundaries are secure before your dogs explore the garden

17 A beautiful bright light is lost ...

Kac

Lisa wrote a moving and enlightened book about Charlie, the wild dog from Romania, who she initially fostered and then adopted, trained, and integrated into her family. Please do not be thinking this is a shameless plug for her book: it isn't. Rather, it is my heartfelt and emotional response to a beautiful canine creature and teacher who I had the privilege to meet and come to know. Charlie is one of the reasons I adopted Talulah and Truffle, and I can honestly say that my life was changed by him in several ways. While Lisa's life was completely impacted, mine was touched by this boy, too.

Her book about Charlie is described as 'The heart-warming true story of how one-eyed Charlie went from traumatized feral dog to joyful family member, bonding with the author, her daughter; making new human and canine friends, and eventually overcoming his fears to settle into his new life.'

All of that is true, and also so much more than that. This is a story about how an amazing creature, beautiful and independent, brought together a community, and taught us lessons about life, change, passion, and extraordinarily deep bonds of true love and friendship.

Charlie, one of the greatest teachers on Earth.

I am honored to count myself among Charlie's friends, and I am grateful to him for sharing his short, sweet life with me.

In her book, Lisa recounts how Charlie arrived at her home in England,

having been transported from a feral and dangerous life in Romania, and describes the daily challenge of getting him to even try to eat out of a dog food bowl. Charlie had already lost an eye, and was adjusting to life through a single lens. Every single day was fraught with new sounds, lights and shadows that frightened him, and sent him, cowering, into hiding. Lisa patiently worked with him, encouraged him, and became very creative in finding new ways to do familiar things that would put Charlie at ease. Day after day, Lisa would recount her struggles and triumphs with this feral boy, and her Facebook community responded like caring aunts and uncles.

There were many times when I thought to myself, 'Is this really worth it?' But I'd quickly come to my senses, read her latest post, and cheer right along with the rest of them about Charlie's newest accomplishment, which may have been simply allowing someone to stroke his face.

As Charlie grew in his domestication, he also anchored himself more deeply in our hearts. His spirit and bravery shone through in the stories and pictures Lisa shared. The huge confusion he sometimes felt was apparent some days, and then a huge smile would cover his face the next. It was like watching a flower bloom into grandeur: Charlie was a very special being.

I corresponded with Lisa as she fell more in love with him each day. It was cause for celebration when she got him to eat his food from a bowl inside her house,and we almost declared a national holiday the day Charlie crawled upstairs and onto her bed. Oh, and there was the time he actually saved her life by waking her daughter in the middle of the night when Lisa was having an asthma attack!

It was the fall of 2014 when I met Charlie face-to-face. I was on a holiday in Edinburgh, and took a train to visit Lisa and her family in Corston, near Bath. Lisa had made so much progress with Charlie by then that greeting him was a joy. I looked at his face, our eyes connected, and we were instant chums. He came to lick my hand and we hugged. From that moment on he was by my side for the evening. I can't tell you what the exact connection was … I like to think that he had read some of my adoring Facebook posts about him. He was so full of joy and happiness it was a delight to spend two evenings with him and his mum. Lisa was thrilled that we played and snuggled, and didn't mind that we had a grand old time with each other.

I had met Lisa's older dog, Skye, a couple of years previously, and seeing him again was also a treat. Skye is loving and friendly, and great fun to be around. We chatted over, under and around the two dogs. Mind you, managing a dinner plate on my lap and trying not to share with Charlie was a balancing feat worthy of circus fame.

When it came time to say goodbye, Charlie and I had a good hug; Skye too, and then I was off to my home in California.

I continued to follow the exploits of this peppery pup on Lisa's Facebook page, and they cheered me no end. But then Charlie began to show signs of recurrent feral behaviour and aggression, and Lisa brought in many experts in Reiki and holistic healing therapies to try and help him. She posted many articles about this syndrome, and her followers were rapt with attention to the daily shifts. Several others in her online community had adopted Romanian dogs, and the teaching was helpful to all struggling with issues. It was possible that changes were taking place in Charlie's brain – maybe as a result of the impact or trauma from the lost eye – but he had thorough medical checks and it turned out there could be a very different reason.

Lisa had discovered that many imported, unsocialized dogs seem to

Charlie and Skye at home in Bath, UK, and my first visit with Charlie.

Skye and Charlie watched the gift-giving with interest.

Kac, Lisa and Marlene with Charlie at his home, and the big 'whoo-hoo' moment when he offered his paw to Kac.

go through a stage of rebellion and aggression after a period of time in a home, and Charlie was undergoing changes, both physical and emotional. Lisa had to keep others away from him for a time as a safety precaution, while she managed the situation and started afresh on helping him cope with the challenges and stresses of his new life. This took months, but she was successful, and Charlie once again blossomed.

Tragically, Charlie died all too soon, after a short mystery illness, on the MRI table where he'd been taken to be scanned after collapsing. He was a brilliant dog, so full of life and spirit; sadly missed and mourned by hundreds of us who followed his daily life. Needless to say, Lisa was heartbroken, and his loss was an earthquake in her world. Skye and Charlie had become best buddies, and Skye also suffered from the loss of Charlie Boy. As so many animal lovers know and have experienced, once a sweet soul touches the deep regions of your heart, there is a chasm of loss and pain when they depart. For a very long time we were all very, very sad, and missed our Charlie.

Unexpectedly, I underwent serious heart surgery later that fall, and the book about Charlie was by my bedside all the time. Whenever I descended into dark days, I would pick up Charlie's book, to read about him and everything he went through, and suddenly my troubles and pain would subside. Once again, Charlie was bringing me joy and comfort in my time of need. Even the memory of his smile can lift my heart. He sure had some power.

Romania has a dog problem. There is not enough funding to spay and neuter dogs. There is also not enough education about the necessity of controlling the dog population in a safe and humane way, so that they are not running loose, resulting in more and more feral dogs who die of abuse and starvation every day.

The problem began in the 1980s, when dictator Nicolae Ceausescu industrialized Romania, forcing people to move to apartment complexes in the city, and share living quarters with other families. Many animals were abandoned as a result of there not being sufficient room for them, and soon the streets were filled with homeless dogs.

The first solution was mass slaughter of the animals and, for more than twenty years, stray dogs were chased, shot, poisoned, hung, burnt to death, or crammed into small boxes and simply left to die of hunger and thirst.

Today, there are still as many stray dogs in Romania, and UK and Romanian animal rights and welfare organizations have stepped in to help by using humane ways to resolve the problem, and by educating the Romanian government and populace. A new animal protection law was launched in Romania in 2008, according to which no healthy animal should be killed. This law did not have the desired effect, unfortunately, as more dogs were then left in overcrowded shelters, to die of disease, injury from fighting, starvation and thirst. Shamefully, dogs are still mistreated today, whilst rescue organisations do what they can to help through sterilisation of the animals and education of the people (http://www.koirienystavat.com/en/dogs-in-romania).

Charlie was one of those fortunate dogs rescued and brought to the UK for fostering and adoption. He was lucky to find Lisa, and have the life he did, albeit shorter than we would have wanted.

There are a great many dogs and other animals in need of help: this is a global problem that we experience in the US, also. Lisa will have many things to add, and I make a point of providing active support to rescuers, shelters, and centers that house dogs for adoption. If we collectively support rescue animals, and give them homes, we could end the killing of healthy dogs in our country, and maybe eradicate the brutal puppy mills that force dogs into a life of dreadful servitude as puppy-making machines. As a civilized society we should know better than to inflict pain, isolation, and cruelty on innocent animals ...

Lisa

I have worked with many troubled dogs, and have taught dog psychology and behaviour for years, but Charlie was my greatest teacher. Helping him through the traumatic transition from the free-ranging, unsocialized animal who was more wolf-like than dog-like (I called him my Wild Soul), to happy, playful, sociable family member led me to draw on every aspect of my knowledge of dogs, thoroughly research the lives of feral dogs, and look at this scary new world through Charlie's eye, instead of from a human perspective. The journey was life-changing for both of us, and Skye's calm, mentoring presence was instrumental in giving Charlie the confidence to take small steps, followed by giant leaps, forward.

In the USA and UK, a heartbreaking number of home-bred dogs end up in shelters and pounds, and the killing figures are shockingly high simply because there are not enough homes available for the perfectly easy, friendly dogs who are taken in because of over-breeding, or abandoned for a variety of reasons. According to the ASPCA, of the 3.9 million dogs admitted to shelters each year, 1.2 million dogs are euthanized. In the UK, the Stray Dog Survey conducted by Dogs Trust charity revealed that of the 103,263 dogs handed in to pounds, 5142 dogs were euthanized.

The figures keep rising, with a great many dogs now being imported from overseas. In the UK, we get dogs sent from Ireland, Spain, Portugal – and a high percentage of these come from Eastern Europe, especially Romania. A lot of dogs are also sent to the USA to find homes. A UK survey published by DEFRA (Department of Food and Agriculture) in January 2016 stated that, in 2015, 93,424 dogs were imported into the UK, of which 33,249 came from Romania, Hungary, Lithuania, Poland, and Ireland. Some of these dogs settled into their new homes. Others did not, and their adopters relinquished them to rescue shelters because they were ill-equipped to cope with the difficulties involved in caring for an unsocialized dog. Rescue worker friends have told me that it feels as if they are trying to empty an ocean with a leaking bucket.

Living with Charlie raised my awareness of the challenges and joys of caring for a dog who had never experienced home life, and led me to investigate the dog importation and puppy farm trades. Seeing how many dogs were in desperate straits all around the world got me thinking about how they could be helped, and the Dog Welfare Alliance (http://www.dogwelfarealliance.com/) was born: a not-for-profit organization I founded with the aim of bringing together dog behaviour professionals, scientists, therapists, rescues, welfare organizations, and members of the public to promote dog welfare, help fund medical care, food and shelter for homeless dogs, and raise awareness of force-free training methods that could help reduce the number of dogs facing a short and uncertain future. I view the Dog Welfare Alliance as Charlie's legacy, along with my book about how I helped him adjust to domestic life.

Through the alliance and my involvement with The Association of INTODogs (a members-only organisation for canine professionals, students of canine behaviour and training, and those working in associated fields), I was among those invited to the House of Commons to discuss with Members of Parliament what could be done to reduce the number of 'dangerous dogs,' clamp down on backyard breeders and puppy mills, and make more public the sad plight of Greyhounds who are bred in large numbers to run for profit, and then simply abandoned or cruelly killed. We still have a long way to go, but, thanks to high-profile celebrity support and the immense power of social media, many of the dog-owning public are stepping on board and paying a great deal of attention to welfare issues.

Bringing in change

There are a number of ways in which each of us can help, and even if we can only do so in a small way, it can still make a major difference, and perhaps even save a life.

Here are just a few of the ways –

- Dogs cannot speak for themselves; they need our voices. Speak out about any form of abuse that you see. Report it to the authorities. Be a protector and champion
- Adopt, don't shop. A horrifyingly high number of puppies are bred in puppy mills – factories for dogs, in which the parents lead wretched lives of constant suffering, with no medical care, no exposure to sunlight or the world outside their cage; totally deprived of gentle touch or kind words. Puppy mill dogs are often called 'crops' because they are bred purely for profit. Don't support this hateful industry: if you want a puppy there are plenty in rescue centres!
- Foster a dog. This enables the rescue shelter to take in another needy dog, and fostering gives opportunities for a dog to be accurately assessed in a variety of situations. It can be bittersweet when a foster dog leaves *your* home for their forever home, but the rewards of fostering are immense, and you are literally saving lives. If you fall deeply in love and can't bear to see your new friend leave, adopt him or her!

- Support your local rescue shelter in any way you can. Perhaps you could make a small but regular donation, or take in spare blankets, vitamins or dog food, or help to walk the dogs in your free time. If you have a skill, such as computing or fundraising, you could offer help in those ways, too
- Spay and neuter your dog, and support spay and neuter organizations in your home country and overseas. A great deal of physical and mental suffering is caused through overbreeding, and in areas where unsocialized feral dogs live it can often be kinder and more compassionate to trap, neuter and release them, so that they can live the free-ranging lives they are perfectly adjusted to without adding to the dog population.
- Explore 'positive' force-free methods and use these with your dog. They strengthen trust and cooperation, and enable relationships to blossom to their full potential. Many dogs in shelters are relinquished because their guardians believed they needed to be the 'alpha,' and used methods of intimidation and punishment on the dog which subsequently broke down all trust.

That last farewell

Charlie was a force for change. My beloved wild soul, my feral boy, touched thousands of hearts and raised awareness of the plight of homeless dogs around the world. His story prompted many people to adopt a dog during the 26 months he was with me – including Kac and her partner who, though previously self-confessed 'cat people,' offered their home, love and commitment to two dogs with sad pasts. Talulah and Truffle are fortunate!

Charlie reminded me of how, tamed though we humans appear to be in our high-tech lives, a part of us from the deep past, our wild selves, still remains, and whispers softly to be expressed.

Sometimes we even pay attention and rediscover that element of pure joy that comes from fully embracing each moment.

Death is always hard to accept. We know that life will come to an end for all of us one day, but until the body succumbs to sickness or old age, we tend to go blithely on, without thinking too much about it. I've cared for many elderly and terminally ill dogs, and have held them and whispered words of love and comfort, tears flowing while they drew that final breath. When Charlie died so suddenly it was a shock. He was a young dog, most likely not more than five or six years old, and I had expected to have years with him. His loss hit hard, and I was astonished and deeply touched that thousands of people, many of whom had never met him but had followed his progress on Facebook, grieved with me and my family. Emails, cards and flowers arrived in abundance, and a friend involved in rescue in Bulgaria sent me a beautiful rose bush called 'Cheerful Charlie' that is in bloom as I write this.

Charlie's ashes and collar rest in a beautiful large porcelain jar in my living room. In life he chose to spend every moment by my side, and it felt right to keep him close after he died. On special dates (the anniversary of his arrival and his passing, and the date we chose for his birthday) I light a scented candle, and focus on the precious memories of our time together. Our loved ones may be gone, but they are never forgotten, and these small rituals bring comfort.

There is a saying by Karen Davison that is quoted often in rescue: "Saving one dog will not change the world, but surely, for that one dog, the world will change forever." Charlie's world changed beyond measure when he was thrust into an environment as alien to him as Mars would be to you and me. He changed my world, too. He changed me. He changed the lives of every dog who was adopted after people met him, read about him on social media, or read my book about our journey together. This book you're reading right at this moment may not have been written if not for Charlie – who captured Kac's heart so strongly that she and her partner adopted two small dogs themselves, and became 'dog people' as well as 'cat people.'

If one dog can cause so much change, just think what ripples may flow from one small act by you!

Summary

- The stray and unwanted dog problem is a global issue
- Many dogs die in shelters through lack of homes
- There are many ways in which *you* can help bring about change
- Saving one dog can transform both your lives

18 The great relapse

Kac

As described in chapter 3, I was elated when Talulah, and – later – Truffle, quickly learned to answer the call of nature to my special word 'Twinkle,' and did not imagine, for an instant, that this handy routine would change. But it did. Overnight.

The move to the new house was fraught with challenges, but the dogs and cats appeared to deal with the tricky times, and settled into the new landscape and territory without too much stress.

I don't know if 'someone' from the canine or feline group accidentally 'went potty' in the other camp's territory, or if there was a close encounter of the strangest kind, but the dogs' house-training went flying out the window like a strong wind on a testy sea.

All of a sudden, the dogs – both of them – decided that the lavender Asian rug was a public restroom for canines. 'Why trouble ourselves to go all the way [15 feet] outside when we can just do our business on this nice, springy, soft rug?' they seemed to think.

I Googled probably a hundred sites in my search for the best deterrent, and something that would eradicate the smell. One company I called quoted me $1100 to clean the 12ft x 9ft rug. I must have invested more than $150 in cleaning agents, rug solutions, specialized

After the move, both dogs needed a refresher course in toileting outside instead of in the house.

'pet' versions of common cleaners, and mail order 'guaranteed' doggie-off. Nothing seemed to work. Talulah and Truffle did use their outside potty area in the mornings, but, for the rest of the day, they used the rug. And I was home and available to let them out. Annoying!

I tried the commercial rug cleaners; I tried mixtures of rubbing alcohol and white vinegar. I tried diluted hydrogen peroxide mixed with all of the above, and it didn't faze them. I even sprinkled cayenne pepper on the soiled areas, but they went anyway, turning the area pink. I blended every concoction that others swore by, all without success.

I tried every trick and approach I could think of. I set up my laptop and camera right next to them, and watched them like a hawk. At the slightest movement, I asked if they wanted to go outside. I tried praise and treats when they did go outside, but then, when I left the room for even a minute, they seemed to think it was 'time to go,' and did so ... repeatedly. I think I was driving *them* crazy hovering like a helicopter parent, but I couldn't help it, I was beside myself with angst. They had been doing so well, but then something triggered this behavior, and I couldn't figure it out.

With my tail between my legs, I emailed Lisa, with H-E-L-P in the subject bar.

Lisa

It's common for potty training to suddenly go up the creek. A number of factors can cause this, just some of which include changes to the usual routine or in the environment, health issues, emotional upsets, rough weather that a dog just doesn't want to go outside in, unnerving noises outside, or an 'accident' indoors that sparks off a new toileting habit.

If a dog begins urinating more frequently, or toileting indoors, and there are no obvious causes such as a change in the physical or emotional environment, I always ask that the dog has a thorough health check to rule out the possibility of a urinary or kidney infection. If there is no underlying health issue, you will need to have a huge cleanup, and then start all over again with house-training.

All members of Kac's household had recently undergone a total change in environment, which was further exacerbated by the upheaval of building work in progress at their new home. This was a stressful situation for the humans, and confusing for the dogs and cats, so it's likely that one little 'accident' when either Talulah or Truffle couldn't get outside in time set the seal on the Asian rug becoming the new toilet (and a nice comfortable place to 'go,' it was, too!).

Another aspect that I found interesting was that Talulah and Truffle were content to go outside in the mornings, but used the Asian rug the rest of the day. What had changed, and what was the payoff for them? Dogs repeat behaviours they find rewarding, and, aside from the rug being a comfortable 'rest room,' it was likely that they got a lot of extra attention for using it. Any attention, even negative feedback such as sighing, chastising, or casting a black look in their direction, would only serve to reinforce the unwanted behaviour.

Bearing in mind that Kac was dealing with a lot of stress, plus was busy making this new house into a home, juggling with fitting her belongings into a slightly smaller space, it's likely that she had less time to devote to the furry family members at that point. The morning routine would involve greetings, breakfast preparations, and going outside, so would be similar to the old routine, but the afternoon routine would most likely have changed from that which Talulah and Truffle knew.

This element could be addressed through Kac setting aside 'dog time' in the afternoons, with the aim of following the routine they had in the previous house as closely as possible. Potty breaks, a walk, short playtimes in-between sorting out all that needed to be done in the house, would help re-establish the routine that the pups felt secure in. Not an easy task when you're surrounded by boxes and chaos, but worth it to have a wee-free, poop-free house.

The first step to take, though (and Kac did this), was to remove all traces of the dogs' elimination scent from the rug. You can buy designated products for this, and a cheaper (and just as effective) method is to scrub the area with a solution of biological (not non-biological) clothes washing powder or liquid. Just avoid bleach, because this doesn't remove the smell for dogs, and can actually attract them to return to that area for toileting.

However, just cleaning the rug isn't enough. When a dog wees, splashes (especially with male dogs) occur in the surrounding areas. They also often step in the wet patch and their paws spread the scent everywhere they go. What Kac needed to do was scrub all of the areas the dogs had access to, to ensure no scent traces remained, and could be construed as new indoor toilets. It's a lot of effort, but well worth it.

It's important to avoid making a fuss when an 'accident' occurs. I suggested that Kac cleaned up quietly without looking at or speaking to the dogs – and, even better – to calmly remove the dogs to another area before the clean-up, when possible. That way, if attention-seeking is one of the causes, the payoff for this ends, which makes the behaviour less rewarding.

It was important to start afresh with toilet training. Regular toilet breaks, and always after food, drinks, naps, and playtimes, are essential, with Kac going outside each time so that she could praise each event, and drop a treat in front of each dog when they performed, so that they knew what they were doing right.

Putting a long house line or leash on each dog so that it trailed after them would make it easier for Kac to step in fast by grabbing the lead and taking them outside as soon as she saw signs of restlessness, sniffing or circling, before the squat or the leg-cock occurred.

Dogs are highly sensitive to our emotional states. Kac told me that she had reached the point of anxiously hovering, watching Talulah and Truffle like a hawk to try and preempt any more assaults on her rug. That tension would have been evident to the dogs, and their response would be to wait until her back was turned, and *then* race to the rug. Some deep, relaxing breaths were needed, and knowing that she could simply pick up the leash as soon as she noticed any signs would help, too.

Kac's relief when the pups stopped toileting indoors was very evident!

Summary

- A lapse in toilet training can be due to health issues, so book a veterinary checkup in the first instance
- Changes in the physical or emotional environment can cause house-training issues
- When an 'accident' occurs, stay calm and avoid paying attention to, or telling off, your dog
- Clean up all areas your dog visits, as well as the area where the 'accident' has occurred, with a designated cleaner or biological washing product
- Start over from the beginning with toilet training, and praise and reward each success

19 Winds just like in the movies

Kac

Southern California is famous for its seasonal winds, called the Santa Anas. They are legendary: hot, dry, extremely turbulent, and the cause of great physical destruction, sometimes. You can't avoid the winds; only endure and survive them.

On the first night of the Santa Ana winds, I put Talulah and Truffle to bed at the regular time, when all seemed well. In the middle of the night, however, the winds picked up. They're so powerful, they remind me of the winds in the movies when a ship is tossed in a stormy sea created by wind machines, fake waves crashing over the deck. We've all watched the sailors scream 'Batten down the hatches, mates,' as the ship rocks back and forth like a tipsy elephant in turbulent rapids. The waves, tall and angry, batter the vulnerable vessel, and we are caught up in the emotion of the scene. Well, it was like that, only worse, and Talulah and Truffle were truly terrified.

In the middle of the night I heard desperate whimpers, and frantic pawing at the crate doors, and ran downstairs to help what I thought might be an injured puppy. Both dogs were emotionally caught up in the storm, and fearful. The motion-detector porch lights were going on and off like searchlights in the high winds; the usually delicate wind chime was sounding off like it was New Year's Eve. The gusty winds were whipping around the house, causing the door jams to whine and squeak, and the potted plants were taking headers and tipping over as they were pummeled by the 80mph-plus winds. Palm trees bent low to the ground, and the swinging chaise lounge was thrown hard against the patio door. The entire house felt like it was under siege.

Poor Talulah was terrified by the powerful 80mph winds ...

Talulah and Truffle were sweating,

... whilst Truffle was completely unfazed, and slept right through them!

panting and scared. I quickly opened their crates, and they leapt into my lap: we sat cuddling for a few moments until they calmed a little. They were both wound up so tight – wiggly and twitchy – circling me twenty times like I was a maypole. I gave them some water. Still, they were wired.

I was beginning to become anxious myself, worried they would never calm themselves. There seemed to be a big difference in how the wind sounded downstairs, where the dogs slept, compared to upstairs, where we slept, so I decided to take them upstairs with me to see if they would settle and relax. I placed them on the bed. The sleeping cats scattered to their usual hiding places – sorry, kitties! Truffle immediately found a soft corner on the big bed and settled down. He was happy and comforted upstairs. Talulah was still wired, though, and it took much petting and cooing in her ear to get her to settle between the big pillows. Eventually, she relaxed and went to sleep. I was so grateful they had both found peace that I didn't mind the snorting and snoring in my ear. We were huddled together, and protected from the storm.

The next morning, Truffle was calmer than Talulah, and took his usual nap, whilst the winds continued to howl. I wrote to Lisa and asked her what to do for the long haul.

Lisa

Back in 1981, the *New York Times* published a piece about the emotional and physical effects of high winds; in particular, winds such as the Santa Ana that Kac and her household were experiencing. A series of experiments at New York University demonstrated that the electrification of the atmosphere during these displays of nature's extraordinary power creates positive ions. In brief, the molecules in the air, such as oxygen and nitrogen, are usually electrically neutral. However, in certain conditions, such as storms and high winds, an electron may be knocked off a molecule, giving it a 'positive' charge. An upsurge of positive ions can affect mood, the state of your health, create feelings of tension and irritability, and reportedly even precipitate a higher-than-usual suicide rate.

Our sensitive dogs experience this change even more powerfully than we do, and the increased electrical charge during very high winds and storms also has an uncomfortable physical effect on them. There are sensory nerves in the hair follicles and dermis, so the piloerection (raising of the hair) that occurs in response to that electrical charge brings with it very unpleasant sensations.

Think of the goosebumps we get when we're afraid, with that attendant creepy sensation of the hairs on your arms standing to attention, then multiply it to imagine your entire body reacting. It's enough to make one shudder just to think of it!

A dog's hearing is far more acute than ours. Our auditory range is between 20 hertz and 20 kilohertz, though few people can hear that full spectrum, and most of us struggle to hear sounds beyond one kilohertz. A dog's hearing extends to 45 kilohertz, up in the ultrasonic range. Imagine being able to hear electrical wires humming in the walls, and the ultrasonic sounds made by rats, and you get an idea of how noisy the world is for your dog, and how overwhelming the rushing and roaring of high winds must be for her.

Noise sensitivity, feelings of distress during exposure to certain sounds, is an issue for most dogs. Sudden noises impact even more strongly on them than on us, and firework season is sheer torture for many dogs.

Then there are the scents to contend with. Our less sophisticated noses pick up on the acrid smells that linger after fireworks have been set off, but we can't detect all the components in the myriad of scents carried in the wind. Watch your dog if you're outside together on a pleasantly breezy day. His nose will be angled skyward, twitching to capture and decipher the vast amount of information that's being carried past in every moment. During very high winds, the overload of information can simply be too much to cope with.

Talulah and Truffle found the high winds terrifying, and no wonder. Their familiar world was caught up in a maelstrom of overwhelming sounds, scents, and sensations, as the trees outside were lashed about, the porch light flashed, and the house creaked and moved around them. Their noses were bombarded by the multitude of scents whirling around, which had travelled great distances. The heightened electric charge and increased presence of positive ions generated sensations of fear, anxiety, and physical discomfort. The only place of safety was close beside their trusted guardians, and Kac's decision to take them upstairs to bed was the right one.

When we know it's going to be noisy, we can begin work beforehand on what behaviourists call desensitization and counter conditioning. This involves exposing our dogs to a very, very low level of the feared noise, and making sure that only good things (treats or games) happen during that process. As the dog shows signs of coping, the volume can be very gradually increased, with great care taken to ensure that no signs of stress – such as flattened ears, rounded eyes, lowered body and tail, and trembling – are being displayed.

This strategy works well for firework season, providing that measures are taken well in advance, but storms and high winds are often unpredictable, and can occur suddenly. It's necessary to keep our dogs safe and help them feel secure.

Safety comes first. Ensure that all points of exit, both indoors and outdoors, are firmly closed (remember to check gates and fences again after all is calm), and that any objects which are likely to move or lose stability are battened down or put away. Stay together in one room with the door closed, so that you know exactly where your furry family members are. If your dog is trying to hide behind or beneath furniture, create a den by draping a throw or sheet over it, and placing something soft to lie on inside. Avoid coaxing your dog to come out – if she feels safest in that place, leave her be.

Until recently, it was thought that, by reassuring our dogs when they are afraid we just compound and reinforce that fear. Fortunately, science has now demonstrated that this simply isn't true. Yes, if you make a huge fuss this could give your dog the message that you're worried, too, and that *could* increase her anxiety, but a calm attitude, gentle strokes and soft words will help her to feel safe, understand that you're taking care of her, and that no harm will come to her. The comfort of your touch, or simply being able to lie close to you, will be a huge relief for a scared dog.

Music can help to soothe noise-sensitive dogs, especially music that's especially designed for them. The *Through a Dog's Ear* series, and Victoria Stilwell's *Canine Noise Phobia* series can be very effective in calming anxious or frightened dogs.

Summary

- Increased positive ions during high winds and storms have profound effects on dogs
- As a dog's sense of smell and hearing are far better than ours, the overload of sensory information can be overwhelming
- Ensuring safety is paramount
- Keep everyone together in one room during powerful acts of nature
- Create a den if your dog tries to hide
- Offer gentle, calm reassurance, and affection
- Music designed especially for dogs can help them to relax

20 Poorly pup

Kac

Only twice in my life with Talulah and Truffle have I had to seek critical veterinary care for them.

One time, Talulah got up on the table, unbeknown to me, and ate a skewer of barbecued chicken kebabs. A few minutes afterwards, I found her hacking and coughing in the corner of the room, shaking her head side-to-side. I was sure she was having a convulsion, and was preparing to rush her to the only emergency vet open on the weekend, forty miles away, when she coughed up half a broken wooden skewer stick. Holy Croatia! We were shocked at what came out of her mouth and throat. I couldn't believe what I saw and immediately had to quell images of internal puncture wounds and leaking inner parts. I held Talulah for a few moments, and was just about to race out the door with her when she looked at me and, basically, told me with her eyes that she was just fine. It was all over.

She was right. I let her run around for a few minutes to be sure, and, after having a drink of water, she curled up and went to sleep. My heart was still beating fast, but somehow I knew I could trust her instincts over my fears. Wasn't it a common thing for dogs to ingest kebabs and cough up the skewer moments later?

Truffle, on the other hand, has had two bad encounters with groomers, and with overly-aggressive trimmers. His first haircut and bath was by a local groomer in our small town. The reviews on Yelp (which publishes crowd-sourced reviews about local businesses) although sparse, were complementary, so I booked an appointment for him there to avoid the ninety-minute round trip to a groomer in the closest city.

The grooming salon was small, but seemed adequate, and, when I dropped him off for his appointment, there were already two lovely small dogs there, waiting in a cage, who seemed happy enough. I found the woman owner to be highly animated and energized, working on the dogs with earplugs in her ears and music playing. I told her that we had just adopted Truffle, a rescue dog, and didn't know much about his history, other than he was a sweet boy. I asked her to please take good care of him, and trim him so that he wasn't so matted and dirty.

She was cheerful about the job, and told me to collect Truffle in two hours.

On my return, a rather seedy-looking guy – who, clearly, hadn't shaved

or bathed in days, – was sitting behind the desk. Truffle was waiting for me in the cage, and he looked anxious to be picked up. When the door was opened, he jumped straight into my arms. I thought he was just glad to see me, but, after I got him home, I noticed that he had a few red marks on him, and what looked like razor cuts. By the next morning he was grimacing and scratching, clearly agitated by the abrasions.

Off to the vet we went, and she prescribed some topical antibiotic cream for Truffle's wounds which she deemed to have come from the trimming tools used. It also looked like he had a burn or two from an instrument that was too hot. Damn! I thought I sensed something wrong. The woman in the shop had been much too energetic, and, having no proof at all, but only pretty good instincts, I suspected she'd been high on something. And the seedy-looking guy didn't add anything positive to the picture, or my opinion of the salon.

I didn't take any action against her: I suppose I didn't think I could. I found another groomer, told them about Truffle's experience, and made them guarantee he would not be harmed or injured in the process of the haircut. And, he wasn't. At least for two years.

I moved location, and looked around for a new groomer, and found one that we liked ... but so did everyone else. She was *always* busy. I couldn't wait two weeks for Truffle to have his haircut so I took him to a local national pet store chain, where I was assured they would be very careful, and he would receive the best care and treatment. By now I knew that Truffle needed hypoallergenic shampoo for his sensitive skin, and that he should have a medium-length trim. The groomer was very sympathetic to my requests, and I provided treats for her to give Truffle in case he became anxious during the process.

When I picked him up, he did the very same thing he had at his first, disastrous trim, jumping into my arms and shaking a little. It hadn't occurred to me that Truffle would be nicked and cut again; especially after being assured he would receive the best care.

Within in a few hours Truffle was rubbing his hindquarters on the floor and rugs, and licking his genital area. I looked to see what was up and noticed redness, and what looked like scrapes. His bum was red from the rubbing he'd done, and little bumps were appearing. He had absolutely been shaved too close in the hindquarters: further inspection revealed he had been nicked in some very tender areas.

I had some antibiotic cream on-hand from the previous injury, but when I tried to apply this, he yelped in pain. With help to hold and comfort him, I gently bathed the area with a dilute solution of hydrogen peroxide to help prevent infection. Poor Truffle whined and grimaced, not enjoying the treatment for a second. After a rest and some treats, I carefully applied some of the antibiotic ointment to the red and swollen areas, and just held him for twenty minutes. When the ointment had penetrated, and he was trying to get free, I used a cool compress on his injured areas to ease the discomfort.

The night did not go well. Truffle was unable to sleep, and neither were we as a result. Marlene and I took turns rocking him and holding him until he finally drifted off.

In the morning I phoned the pet store and spoke with the groomer. She immediately passed me on to her manager, who told me to take Truffle to their vet, whilst trying to claim it was not their fault. Hardly!

After a full examination the vet confirmed that Truffle's injuries were caused by the clipping, and recommended the same ointment I had at home, plus diapers to protect the area as it healed. Truffle was clearly in pain. We

Truffle required antibiotics, pain relief, and five days to recuperate after a careless groomer nicked him in three places, causing infection.

returned to the national pet store chain to speak with the manager, and confirm that Truffle's injuries were directly related to the grooming he had received there.

For the next two days Truffle lay on the sofa, wrapped in his favorite blanket. He could not, or would not, urinate or defecate, and wasn't even interested in eating. Truffle? No food? Something was terribly wrong. I tried everything to tempt him, and finally got him to eat a little fresh-roasted chicken breast. He didn't want any water, however, and even refused his favorite treats.

We waited forty-eight hours before we called the vet again, worried about his lack of elimination. Mightn't he explode? The vet examined Truffle again, and said she thought he might have pancreatitis, which an x-ray could confirm. Again, here was a definite attempt to place responsibility for Truffle's wounds elsewhere. The vet did not have x-ray facilities at her clinic, and sent us off to a vet who did.

Truffle and I waited for our appointment. He was patient, yet clearly uncomfortable. This vet, Dr Amy, was incredible. She took a look at the wounded areas, absolutely confirmed that they had been inflicted during grooming, and said that he didn't need a costly x-ray; he just needed to get that scratchy diaper off, take some pain medication, and wear an inflatable collar so that he wouldn't lick the sores

Within twenty-four hours Truffle was so much better, and able to resume his pattern of healthy elimination because he no longer hurt, thanks to the pain meds. He was also able to sleep through the night for the first time in three nights – and

so were we. It took a full eight days for him to recover to the point that he didn't need pain medication, but what a week that was, with Truffle in pain, frightened, and unhappy. Our week was taken up by multiple vet visits, constant care for him, and the expense associated with the recovery process. And all because one groomer didn't have the requisite skills, or heed our advice.

The end result of this fiasco was that I purchased a dog grooming clipper tool; a quality one. I have visited several sites on YouTube already, and watched many good videos on home grooming. I have decided that I will never subject Truffle to another groomer unless I can be with him the entire time and supervise every move. Otherwise, I'll clip him and groom him myself.

Talulah, on the other hand, doesn't require clipping, so there's no problem with her going to the groomer, and no risk of injury. I'm sure I'll master grooming in good time, with Truffle as my guinea pig. There may be the odd occasion when his coat is a little lopsided, granted, but his hair does grow very quickly, so that should save him embarrassment at the dog park …

Lisa

Dogs are opportunist feeders, who, in the distant past, had to fend for themselves. Meals tended to fall within the feast or famine categories; they would gorge

Truffle was happier, and fared much better, with Talulah's groomer.

on whatever food was available, not knowing when the next meal would be. Even though our family dogs have regular mealtimes (and I know dogs who eat better quality food than their guardians!), that instinct to grab what they can is still part of their nature.

It's hardly surprising, then, that when Talulah saw an opportunity, she helped herself to the chicken kebab, even though it wasn't good for her. She was very lucky to avoid serious injury, and it must have been terrifying for Kac to witness what happened. Talulah came to no lasting harm, thank goodness, but other dogs are less fortunate. Cooked bones (especially chicken bones) can be lethal because they splinter, so it's a good idea to make sure that all food is shut safely away from keen noses.

One of my dogs really took the prize for food thievery. Carnie was given to us by a farmer when we lived in the country in Ireland. He was a mixed breed, no doubt with a dash of Collie somewhere in his genes, and he grew from a fluffy, bear-like pup into a magnificent black, wolf-like dog. He was mischievous and funny; very loving, and would spend hours romping and playing in our huge garden with one of my sons.

Our water came from a well in the garden, and was pumped into the house. A huge Aga range in our kitchen provided cooking facilities, and heated the water for our radiators, as well as taps. Carnie loved to hide quietly beneath the kitchen table, ready to snaffle up any food that came out of the Aga oven if I was foolish enough to turn my back for just a second.

One day, we were expecting guests, and I had made a big batch of vol au vents. I put one tray on the table, turned to fetch the second tray from the oven, and looked back to find an empty table. Sure enough, there was Carnie, under the table, huffing quietly (you can imagine how hot that pastry was!), and waiting for the second serving.

More on groomers

Truffle's experiences of being badly hurt by two groomers was agonizing for Kac and her partner, as well as immensely painful for poor little Truffle. It's hard to leave your dog with a stranger, and it's so important that you can trust that person to take good care of him. If you can find a groomer who will let you stay with your dog throughout the session, that is ideal, but many groomers won't agree to this.

In Chapter 9 I included a section on how to choose a groomer, which contains a checklist of what to look for, and questions to ask before you leave your precious dog in the hands of a stranger.

The grooming industry isn't regulated in all states, which means that anyone can set up a grooming business without proper training. Qualifications can (and should) be acquired, however, so ask for proof of these if you can't see any framed certificates on the salon wall. Ask friends to recommend good groomers (and to share negative experiences so that you know which to avoid). Visit the websites of recognized pet professional organizations, as these should have a list of their members so that you can find a professional in your area. Three well-respected organizations are: The National Dog Groomers Association of America, the Pet Industry Federation, and The British Dog Groomers' Association.

If you're willing to take a course or to study through videos and books, home grooming is a good option for dogs whose coats aren't very high maintenance. It's less stressful for your dog (and for you), and if he looks a little lopsided the first time – well, fur grows back quickly! Some breeds and mixed breeds do need professional hair care, but it's better to wait a couple of weeks for an appointment with someone you trust than go elsewhere and regret it.

Please visit your vet if your dog looks uncomfortable and unhappy after being groomed, especially if you notice nicked

or sore areas, or shaving rash. If your dog is sore or unwell after grooming, tell the groomer, take photos, put it in writing for formal backup, and speak to the person in charge of the facility. If the groomer is a member of a professional organization, inform the organization, too. These organizations expect a certain standard from their members, and rely on the public for feedback on whether these are being met.

Nursing a sick dog

It's deeply upsetting when your dog is sick. As he can't always show you where it hurts, the first port of call should be your veterinarian if he shows any sign of pain or discomfort. When a dog is seriously ill this can take a huge emotional and financial toll on you. It's vital to take out pet insurance as soon as your dog comes to you because even a healthy dog can suddenly become ill, whether through an infection, tummy upset, or the onset of a medical condition. It can be difficult to get insurance if your dog has what is called a 'pre-existing medical condition' because you have applied for this *after* your dog has been unwell.

It can be hard to stay strong and positive when your dog is sick, but our dogs sense our emotional states keenly, so it can help to draw on any support that's available to avoid the possibility of distressing him further. Your partner, family or friends can help simply by listening to your concerns, or giving much-needed hugs. A friend arriving with a ready-made meal is a true gift when you're too busy looking after your dog to have chance to cook. Plus, if you're worried, you may not feel like eating until something tasty is presented to you.

A sick dog may lose interest in food and water. You could offer frequent small, light meals, such as a little cooked chicken breast mixed with boiled rice, as this causes less strain on the digestion. The digestive system tends to shut down during illness, while the body focuses on areas that need more attention. If he refuses to eat, just offer a little occasionally and try not to get upset if he declines. If the veterinarian has recommended a special diet, feed him separately to your other dogs, and make sure all food bowls are removed when your dogs have eaten, so that each has the correct diet.

Water is important, because dehydration will make your dog feel worse, but some dogs find it hard to drink from a bowl during illness. You could offer small quantities of water in a saucer, instead, because the reduced size and shallow shape makes drinking easier. One of my elderly foster dogs who was sent to me for terminal care would only drink from the palm of my hand during his final days. That combination of easy accessibility (I could hold my hand just below his muzzle so that no cold hard edges from a bowl touched him) and my scent combined with the water was comforting for him.

If your dog is having difficulty sleeping, you could either take him into your bed so he can snuggle with you, or set up a temporary bed in the living room to keep him company, and have cat-naps when he's resting quietly.

Any course of prescribed medication will need to be completed. Some dogs dislike taking medication, but you can hide it in something tasty. Fortunately, all of my dogs have loved cheese, so when one is on medication I keep some grated cheese in the refrigerator, and squeeze a small amount into a soft ball, with the tablets hidden in the centre. Some dogs prefer pâté or a little chicken or ham to help the medicine go down.

Your company and love will have its own healing influence. Keep the environment quiet and calm, as this will help you as well as your dog, and call your vet if his condition worsens or you have any concerns.

Dogs are amazingly resilient,

Truffle shows his contentment by cuddling with his Happiness Pillow.

and, unless they have a serious medical condition, will bounce back quickly once healing begins. It can help to keep life low-key for a few days after he's begun to recover, even if he seems to be raring to go. Light meals, plenty of water, and shorter walks and playtimes, just temporarily, will help him rebuild his strength.

Summary

- Dogs are opportunist feeders
- Choose your groomer based on recommendation and qualifications
- You can learn to groom a low-maintenance dog yourself
- Pet insurance is essential in case of future illness
- Dogs sense our emotions, so our worry and distress can upset them
- Offer small, light, frequent meals to a sick dog
- A saucer may be easier for a sick dog to drink from
- Dogs are surprisingly resilient

21 When the student is ready, the teachers will appear

Kac

I may have been brainwashed by the dog food commercials on TV, but I admit to having had a fantasy about adopting a dog. I've had many decades of experience of adopting adorable, irresistible kittens, and then watching them grow into cats. I love my cats, but they do transmute as they grow up.

I believed that adopting a dog would be a one-way street of total joy and happiness; to feed, walk, and endlessly enjoy the love and affection of a dog. I was extremely confident that I was more than qualified to handle this.

A dog would love me, obey me, go to sleep on command, and be a perfect little companion. My dog would come complete with built-in training radar, and behave perfectly from day one. I believed those commercials. After all, I'd adopted many cats in my lifetime – twelve, to be exact – and had a good track record with felines. I was riding high in my confidence convertible, with ample self-assurance blowing in the wind.

Then I met my match.

I could not know, of course, that Talulah and Truffle would become two of my greatest teachers and best friends. Talulah stole my heart first, and we were off on an adventure and steep learning curve which turned out to be largely mine. She was the first to show me that dogs have feelings, fears, preferences, distinct personalities, and specific physical and emotional needs, and they certainly don't arrive with manuals, built-in training, or warranties.

Truffle showed me how sensitive dogs can be to both positive affection and negative words. I learned from him that the most subtle of emotionally tinged communications, even facial or bodily gestures, are picked up by a dog at his or her deepest core. If I was sending a positive vibe, Truffle picked it up and responded with a happy and joyous heart; his inner joy apparent from his wagging tail and sparkling eyes. If I was displeased about something or corrected him, he would visibly sag, and fold into a heap of shame. I was astonished. I couldn't believe that a dog could react to my output in this way. Talulah was braver and bolder than Truffle, but I also began to understand just how sensitive these canine creatures truly are.

I quickly learned that it was I who would mold and structure my relationship with my dogs, because I could see the results of my every action and attitude in their sweet faces. There wasn't a bone in my body that wanted to hurt them or their feelings, but I did have to make some personality adjustments in myself when I needed to correct their behavior, or when I spotted a potty mistake on my new rug.

Ours became a partnership filled with ups and downs, successes and failures, which took time to build and develop, and how my dogs fared and felt was totally and completely dependent on how I handled any given situation. If I reacted with an emotional outburst, their little faces would show their hurt feelings or confusion. They were instant Karma, if you will, and I received lesson after lesson about human behavior – namely, mine.

When Truffle was injured by a careless groomer and required round-the-clock monitoring of his cuts and abrasions, I watched him manage his pain to make me happy. He struggled to be a happy dog, but I could see the pain in his eyes. Both Talulah and Truffle are happiest when we are all in sync. They don't like to see me mad, sad or fearful, and I don't like to see them that way, either, so we work with each other to brave the challenges of daily life: gardeners, mowers, workers, cars, noises, and interruptions. We have come up with special codes that signal our comfort or distress. They probably know me better than I know myself.

Talulah is a natural guard dog, even though she's a serious lightweight. She barks to scare off would-be intruders, even if it is only a neighbor passing by, or the ever-familiar mail delivery person. At first I yelled at her: "No barking!" Then I bought an air can that makes a hissing noise to try and change her behavior. Lisa pointed out that aversive methods such as air cans can create fear and result in loss of trust, and suggested that, instead, I give Talulah treats the moment she pauses for breath in-between barks, to reinforce quiet and reassure her that things are alright. Now, when she feels the urge to bark, she looks over at me first. I mutter a quiet 'no' and she jumps off her perch and comes over to me for a hug. I've even heard her stifle a deep guttural growl, before looking at me, as if to say, 'I'm really trying, mommy, but oh, you don't know how much I want to growl at them.' It always makes me laugh.

Truffle isn't a first responder; he only barks if Talulah does, then looks a little confused about why he's barked at all. He's a mellower fellow, and goes along for the ride.

They have taught me about sweet, quiet moments. Talulah and Truffle are both capable of running up a storm, jumping through hoops, and then settling down to watch some television. They can go from being in 'high' to 'cuddle' mode in an instant, and appear to thoroughly enjoy the soft, simple moments where there's no one else in the world but us. A gentle lick to my hand or chin lets me know they are content and happy to be with me at that exact moment, and, for me, there is no space to worry about bills, the state of the world, or assorted human stresses or problems. We just *are*, and that's enough for us. I learned from them how to be in the moment. It's a process, and we're getting there.

They aren't afraid to show gratitude, which has taught me to make sure I tell people how grateful I am for what they do for me. With everyone, from those I hire to the people who service my car, I make a special effort to let them know how much I appreciate them. It's the same as wagging a tail like the puppies do. I've learned to wag my tail more frequently, even in the most unexpected places, and I've seen how it brightens someone's day. Just like my pups brighten mine.

I have learned to stop and recognize special moments, even when I am on a deadline or late to an appointment. They have taught me about loving moments when they curl up under my desk, or sit at my feet, wearing kind eyes and a doggie smile. They have taught me that it's never too early or too late for a spontaneous kind smile or a hug. It only takes a moment for that special connection to occur, and it revives the soul.

Truffle even knows when it's 5:30pm, and time for the nightly news on TV. He finds me wherever I am and reminds me, with a paw on my lap, that it's time for us to have our half-hour before we start dinner or finish up with work. He's as regular as Big Ben, and never misses a day. Most of the time we stop what we're doing and take that half-hour for news, and just be with each other.

Years ago I took a lesson from *The One Minute Manager* by Kenneth H Blanchard and Spencer Johnson. The book alleges we could be more effective as a manager if we spend just sixty seconds connecting with colleagues, and focusing on their issues, challenges or needs. I had tried the methods at work, and the results were indeed very effective: co-workers felt heard, appreciated, and acknowledged by the one minute of concentrated attention and focus I gave them.

It turns out that this method works in many different situations, and for all kinds of relationships. Instead of being preoccupied, rushed, crazed, or in a hurry to move onto the next task, the very act of giving my partner, child, co-worker or friend one minute of unadulterated focus seemed to be a healing elixir. People felt satisfied when they had their minute of time and attention.

I decided to try the practice on my dogs. When they became demanding, disruptive, or, in my opinion, annoying, I would simply stop what I was doing and give them one minute of total attention. The minute didn't disrupt my work, it was a break, and I benefited from the connection as much as the pups. I was amazed at how satisfied they were after the minute of affection and attention: they settled down, went back to their naps, or succeeded in alerting me to a much-needed potty break.

Another thing they taught me is that kindness at any level is received and returned as kindness; gentleness, too. It's never a bad choice to be kind or gentle. When I am gentle with them, they lap it up like warm milk, and when I am kind to them with extra affection or attention, they take it in like meadow-fresh air. Kindness is an elixir in our house, and when I am on the receiving end, I feel it.

They taught me what it's like to be loved by another being, even if I am tired, look like heck on a stick, or have gained a few lumpy pounds. They like me just the way I am, no matter in what condition, or what my opinion of myself is that day. I have learned to overlook some external appearances, and see more of the inner truth. I'm not totally there yet, but they give me the strength to carry on.

I've learned that a dog can bring out the best, or the worst, in people. The dog remains the constant, but the person reveals their true colors and shades of light or darkness in the presence of a dog. They are mirrors, and we can learn a lot if we are willing to see ourselves through their eyes.

There are many more lessons for me to learn from my furry little companions, and I am excited for what is yet to come. I can always replace a rug, a sofa, or a favorite show, but I can never replace the love in their hearts for me, and the way I feel when they wag their tails when they see me. Thank you, Talulah and Truffle. My heart is yours.

Lisa

The story of Kac's journey with Talulah and Truffle beautifully illustrates how precious the bond is between us and our dogs, and how much we can learn from our canine friends if we're willing to pay attention. It's a partnership, and although it's our responsibility to teach our dogs how to live comfortably, safely, and happily in our world, the lessons are shared in both directions because dogs also teach us a great deal about ourselves. They can change our perspective on life and relationships with others, and their sensitivity to our emotional states, facial expressions, and tone of voice can alert us to how we express ourselves; prompting us to consider whether we could find more appropriate or healthy ways in which to do this.

Dogs don't make a secret of how they're feeling in any given moment. They respond instantly and spontaneously, so if you're willing to learn to understand their body language, it can be a revelation and source of deep joy to communicate at a profound level with another species. When we love our dogs, their pain becomes our pain; we suffer with them, just as their happiness lifts our spirits and lights up our days.

Understanding and meeting the physical, emotional, and mental needs of our dogs enables us to connect with them in profound ways, and nurtures these very special inter-species relationships. Linda Michaels MA, Psychology has created a wonderful system and graphic called the Hierarchy of Dog Needs™ which gives clear guidelines on these. You can find out about this at http://www.dogpsychologistoncall.com/hierarchy-of-dog-needs-tm/.

A family member

Your dog is a member of your family, a social creature who, like you, needs to feel loved and accepted within his group. He looks to you to be his wise guide through the complexities of modern life, and, for this reason, it's important that when you bring home a dog you make sure that every member of your family is involved with his training. Consistency is vital, because if one person treats him differently to the rest, he'll become confused and slow to learn. Working together with a dog can help you form ever closer bonds with your loved ones, too.

It can help to have a family meeting before you take in a dog, to give all of you the opportunity to discuss what you hope for, what your expectations are (and whether these are reasonable), and how you will go about teaching your dog to toilet outside, walk on leash, and learn all of the things you will be aiming to teach him. Having a common goal means that everyone is on the same page; you'll be consistent in how you interact with your dog, and this will give him the best possible start, setting the foundations for a long and happy relationship.

All you need is ....

Love – yes, of course! And also patience, and a willingness to learn from your mistakes and those your dog makes. Staying power helps, especially if you have a puppy who is going through the challenges that adolescence brings. Sensitivity to your dog's needs and respect for his innate doggy nature mean that you give him free rein to do what dogs love to do – run freely in safe spaces, roll in sweet-smelling grass, paddle in streams or the sea on hot days, have a good social life that includes other dogs as well as people, and provide nutritious food that contributes to maintaining his health.

Life with a dog (or dogs) is tremendously rewarding and enriching. Our furry friends encourage us to get out and exercise more, to smell the flowers and appreciate our surroundings. Dogs are great ice-breakers, and many friendships begin after a wagging tail has set a conversation rolling. Dogs can make us laugh and enable us to do silly things that may otherwise make us feel self-conscious (I've run in the opposite direction through fields, waving my arms in the air and calling out 'Yippedy doo-da') to encourage a dog to follow me while teaching recall. It works beautifully, but it

can bring strange looks from observers.

Have fun with your dog. Dogs – even old dogs – know how to have fun, and this is something we tend to let slip away once we're dealing with the responsibilities of adulthood. We take ourselves too seriously, and dogs can teach us to lighten up.

Make the most of every moment, and appreciate the dog hair on the couch and the warm, damp nose resting in your lap. Laugh about those muddy pawprint works of art on the kitchen floor, because they're evidence that your dog's been out there having a great time, and what's a little mud and water between friends? Drink in his love and devotion, and offer him the same in return. You'll be so glad you did!

22 Are you ready to raise a dog?

Take our brief quiz to determine whether you are ready to adopt a dog of your own!

Yes	No	
O	O	Do you have enough free time to take care of a dog?
O	O	If you have other companion animals, how well do you think they will integrate?
O	O	Can you afford pet insurance? The average cost of taking care of a dog can be quite high, even without taking into account medical treatment
O	O	Do you have someone you trust to take care of him when you're away?
O	O	Are you able to walk your dog at least twice a day?
O	O	Are you considering a puppy or a mature dog?
O	O	Puppies need training and attention. Can you set aside enough time for that?
O	O	Mature dogs may require extra veterinary care. Can you afford to give her a comfortable retirement?
O	O	Do you have a garden or access to a place for a dog to potty?
O	O	Are you willing to share your furniture with a dog?
O	O	Praise goes a long way with dogs. Do you have the patience to help him learn from mistakes in a positive, compassionate way?
O	O	Have you considered what diet will provide the best nutrition for your dog?
O	O	Will you take your dog on vacation with you?
O	O	Are there other dogs in your neighborhood? Are they friendly or aggressive?
O	O	Will you train your dog yourself, or get help from a licensed trainer or school?
O	O	Have you taken the time to learn about dog body language and what it means?
O	O	Is everyone in your family enthusiastic and committed to the idea of a dog?
O	O	Have you checked out qualified, accredited, force-free trainers and behaviorists in your area?
O	O	Have you spoken with your veterinary surgeon about health checks, vaccinations, and microchipping?
O	O	Are you willing to remain committed even through challenging times and for many years?

Index